THE CARTOON GUIDE TO CHEMISTRY

ALSO BY LARRY GONICK

THE CARTOON HISTORY OF THE UNIVERSE, VOLUMES 1-7

THE CARTOON HISTORY OF THE UNIVERSE II, VOLUMES 8-13

THE CARTOON HISTORY OF THE UNIVERSE III, VOLUMES 14-19

THE CARTOON HISTORY OF THE UNITED STATES

THE CARTOON GUIDE TO THE COMPUTER

THE CARTOON GUIDE TO THE ENVIRONMENT (WITH ALICE OUTWATER)

THE CARTOON GUIDE TO GENETICS (WITH MARK WHEELIS)

THE CARTOON GUIDE TO (NON)COMMUNICATION

THE CARTOON GUIDE TO PHYSICS (WITH ART HUFFMAN)

THE CARTOON GUIDE TO SEX (WITH CHRISTINE DEVAULT)

THE CARTOON GUIDE TO STATISTICS (WITH WOOLLCOTT SMITH)

THE CARTOON GUIDE TO
CHEMISTRY

LARRY GONICK
& CRAIG CRIDDLE

HarperResource

An Imprint of HarperCollinsPublishers

HarperCollins books may be purchased for educational, business, or sales promotional use. For information please write: Special Markets Department, HarperCollins Publishers, Inc., 10 East 53rd Street, New York, NY 10022.

FIRST EDITION

Library of Congress Cataloging-in-Publication Data has been applied for.

ISBN 0-06-093677-0

05 06 07 08 09 ❖/RRD 10 9 8 7 6 5 4 3 2

CONTENTS

CHAPTER 1 1
    HIDDEN INGREDIENTS

CHAPTER 2 17
    MATTER BECOMES ELECTRIC

CHAPTER 3 45
    TOGETHERNESS

CHAPTER 4 67
    CHEMICAL REACTIONS

CHAPTER 5 85
    HEAT OF REACTION

CHAPTER 6 105
    MATTER IN A STATE

CHAPTER 7 129
    SOLUTIONS

CHAPTER 8 141
    REACTION RATE AND EQUILIBRIUM

CHAPTER 9 165
    ACID BASICS

CHAPTER 10 191
    CHEMICAL THERMODYNAMICS

CHAPTER 11 209
    ELECTROCHEMISTRY

CHAPTER 12 227
    ORGANIC CHEMISTRY

APPENDIX 243
    USING LOGARITHMS

INDEX 245

TO

DEON CRIDDLE,
WHO ALWAYS HAD TIME TO HELP
HIS SON WITH SCIENCE FAIRS

AND

THE MEMORY OF EMANUEL GONICK AND
OTTO GOLDSCHMID, CHEMISTS BOTH

THE CARTOONIST WOULD LIKE TO THANK HIS ASSISTANT, HEMENG "MOMO" ZHOU, WITHOUT WHOSE COMPUTER SKILLS, ARTISTIC ABILITY, AND GOOD HUMOR THIS BOOK WOULD HAVE TAKEN FOREVER...

Chapter 1
Hidden Ingredients

THE VERY FIRST CHEMICAL REACTION TO IMPRESS OUR ANCESTORS WAS **FIRE**.

PERSONALLY, I WAS FIRST STRUCK BY DECAYING MEAT!

WELL, MAYBE FIRE WAS THE **SECOND** ONE, THEN...

WHAT ABOUT PHOSPHORESCENCE? PHOTOSYNTHESIS?

PHLATULENCE?

PHFFTH

ANYWAY, FIRE WAS **REALLY, REALLY IMPORTANT**.

FIRE—AND THOSE OTHER PROCESSES—REVEALED **HIDDEN FEATURES OF MATTER.** IF YOU HEAT A PIECE OF WOOD, ALL YOU GET IS A HOT PIECE OF WOOD, AT FIRST... BUT SUDDENLY, AT SOME POINT, THE WOOD BURSTS INTO FLAME. WHERE DID **THAT** COME FROM?

CHEMISTRY IS THE SCIENCE THAT ANSWERS THAT QUESTION, AND **CHEMICAL REACTIONS** ARE THE STRANGE TRANSFORMATIONS THAT REVEAL MATTER'S **HIDDEN PROPERTIES.**

CHEMISTRY IS A SCIENCE ABOUT THE OCCULT, THE HIDDEN, THE INVISIBLE. NO WONDER IT TOOK SO LONG FOR CHEMICAL SECRETS TO COME OUT... AND IT ALL STARTED WITH **FIRE.**

PROBABLY THE BEST THING ABOUT FIRE WAS THAT IT COULD BE USED TO CONTROL **OTHER** CHEMICAL REACTIONS: COOKING, FOR EXAMPLE!

MM... BIG IMPROVEMENT OVER PUTRID...

YOU KNOW HOW SCIENTISTS ARE: IF THEY CAN COOK ONE THING, THEY'LL COOK ANOTHER. PRETTY SOON, THEY WERE COOKING ROCKS.

YOU'RE JUST TRYING TO PROVE MEN CAN'T COOK, AREN'T YOU?

SOUNDS CRAZY, BUT ONE OF THOSE GREEN, CRUMBLY ROCKS MELTED, CHANGED, AND BECAME AN ORANGE LIQUID THAT COOLED INTO SHINY, METALLIC **COPPER.**

NEEDS SOME KIND OF SAUCE, THOUGH!

THIS ENCOURAGED THEM TO SMELT RED ROCKS INTO IRON... BAKE MUD INTO BRICKS... SAUTE FAT AND ASHES INTO SOAP... AND (WITHOUT FIRE) TO CURDLE MILK INTO YOGURT... FERMENT GRAIN INTO BEER... AND CABBAGE INTO KIMCHEE. THE NEXT THING YOU KNEW, CHEMISTRY HAD CAUSED **CIVILIZATION!**

WHAT ACCOUNTS FOR MATTER'S SECRETS? THE ANCIENT GREEKS CAME UP WITH AT LEAST THREE DIFFERENT THEORIES.

THE **ATOMISTS,** LED BY **DEMOCRITUS,** THOUGHT THAT MATTER WAS MADE OUT OF TINY, INDIVISIBLE PARTICLES, OR **ATOMS** (A-TOM = "NO CUT"). IF YOU CUT AND CUT AND CUT AND CUT, THEY REASONED, THE PROCESS HAD TO STOP SOMEWHERE.

IF OBJECTS HAD INFINITELY MANY PIECES, THEN EVERYTHING WOULD TAKE FOREVER!

INSTEAD OF ONLY SEEMING THAT WAY...

ANOTHER PHILOSOPHER, **HERACLITUS,** SUGGESTED THAT EVERYTHING WAS MADE OUT OF **FIRE.**

YES... WE'RE GETTING WARM NOW...

BUT ATOMS COULDN'T BE SEEN, AND... FIRE? I MEAN, REALLY! THE GREAT **ARISTOTLE** ANNOUNCED THAT THERE WERE REALLY **FOUR ELEMENTS,** OR BASIC SUBSTANCES, FROM WHICH ALL ELSE WAS COMPOSED. THESE WERE **AIR, EARTH, FIRE,** AND **WATER.** OTHER STUFF, HE OPINED, WAS A BLEND OF THESE FOUR.

MAKES SENSE TO ME!

OF THE THREE IDEAS, FOR SOME REASON, IT WAS ARISTOTLE'S THAT MOST INFLUENCED MEDIEVAL SCIENCE. IT WAS SO **OPTIMISTIC!** IF EVERYTHING WAS A MIXTURE OF FOUR ELEMENTS, THEN YOU SHOULD BE ABLE TO TURN ANYTHING INTO ANYTHING ELSE JUST BY TWEAKING THE INGREDIENTS!

LEAD INTO GOLD, FOR EXAMPLE...

THIS HOPELESS QUEST WAS TAKEN UP IN PERSIA BY **JABIR** (EIGHTH CENTURY) AND **AL-RAZI** (TENTH CENTURY), WHO INVENTED ALL SORTS OF USEFUL LAB EQUIPMENT AND PROCEDURES IN THE PROCESS. THIS PROVES YOU CAN MAKE TREMENDOUS PRACTICAL PROGRESS WITH STUPID IDEAS.

ANY GOLD YET?

LET'S REDEFINE OUR GOALS...

MEDIEVAL EUROPE BORROWED THE ISLAMIC SCIENCE—AND ITS NAME, **ALCHEMY** (="THE CHEMISTRY" IN ARABIC)—AND ITS HUNGER FOR TRANSMUTED GOLD. THE GERMAN ALCHEMIST **HENNIG BRAND,** FOR EXAMPLE, TRIED TO MAKE GOLD BY DISTILLING 60 BUCKETS OF URINE.

WHERE DO YOU EVEN **GET** 60 BUCKETS OF URINE?

IN THE END, BRAND'S EQUIPMENT GLOWED IN THE DARK. HE HAD DISCOVERED PHOSPHORUS—BUT NO GOLD...

DESPITE THEIR WILDER SPECULATIONS, THE AL-CHEMISTS ACCOMPLISHED A LOT IN THE LAB: THEY PERFECTED DISTILLATION, FILTRATION, TITRATION, ETC... THEY ADVANCED GLASSMAKING, METAL-LURGY, EXPLOSIVES, CORROSIVES... AND THEY INVENTED "FORTIFIED WINE," I.E., HARD LIQUOR...

AH! INNER ALCHEMY!

BUT THEIR LAB TECHNIQUE MISSED ONE BIG THING: THEY FAILED TO COLLECT **GASES.** IF A REACTION CONSUMED GAS, THE ALCHEMISTS HAD NO WAY OF KNOWING. IF IT GAVE OFF GAS, THEY LET IT ESCAPE.

THE FASTER THE BETTER, PREFERABLY!

THIS MEANT THEY COULD NEVER FULLY ACCOUNT FOR THE **IN-GREDIENTS** OR **PRODUCTS** OF CHEMICAL REACTIONS.

THE MODERN STUDY OF GASES OR "AIRS" BEGAN IN THE 1600s, WITH SOME INVESTIGATIONS INTO THE EFFECTS OF AIR PRESSURE. CONSIDER THIS DEMONSTRATION BY **OTTO VON GUERICKE** (1602–1686).

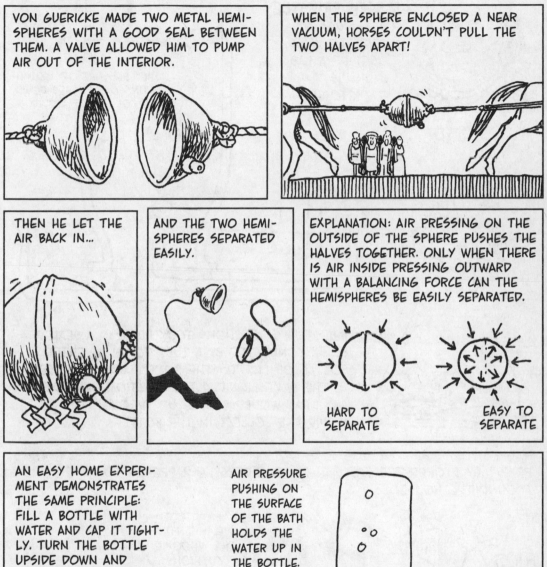

VON GUERICKE MADE TWO METAL HEMISPHERES WITH A GOOD SEAL BETWEEN THEM. A VALVE ALLOWED HIM TO PUMP AIR OUT OF THE INTERIOR.

WHEN THE SPHERE ENCLOSED A NEAR VACUUM, HORSES COULDN'T PULL THE TWO HALVES APART!

THEN HE LET THE AIR BACK IN...

AND THE TWO HEMISPHERES SEPARATED EASILY.

EXPLANATION: AIR PRESSING ON THE OUTSIDE OF THE SPHERE PUSHES THE HALVES TOGETHER. ONLY WHEN THERE IS AIR INSIDE PRESSING OUTWARD WITH A BALANCING FORCE CAN THE HEMISPHERES BE EASILY SEPARATED.

HARD TO SEPARATE

EASY TO SEPARATE

AN EASY HOME EXPERIMENT DEMONSTRATES THE SAME PRINCIPLE: FILL A BOTTLE WITH WATER AND CAP IT TIGHTLY. TURN THE BOTTLE UPSIDE DOWN AND IMMERSE THE CAPPED END IN A WATER BATH. (THE KITCHEN SINK WILL DO.) REMOVE THE CAP UNDER WATER. THE BOTTLE REMAINS FULL.

AIR PRESSURE PUSHING ON THE SURFACE OF THE BATH HOLDS THE WATER UP IN THE BOTTLE.

THIS UPSIDE-DOWN BOTTLE BECAME A GAS COLLECTOR IN THE HANDS OF **JOSEPH PRIESTLEY** (1733–1804), A MINISTER WHO SET UP A LAB IN HIS KITCHEN.

THE PRESSURE OF ACCUMULATING GAS PUSHES DOWN THE COLUMN OF LIQUID.

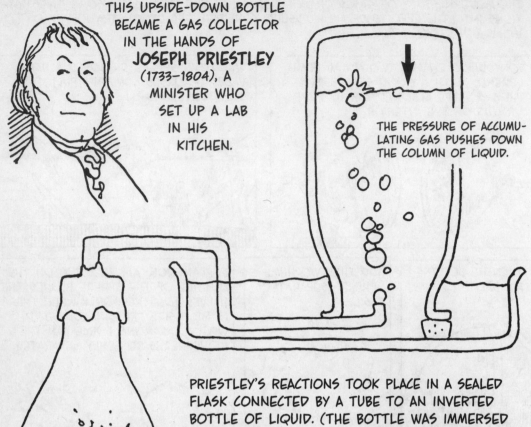

PRIESTLEY'S REACTIONS TOOK PLACE IN A SEALED FLASK CONNECTED BY A TUBE TO AN INVERTED BOTTLE OF LIQUID. (THE BOTTLE WAS IMMERSED IN THE SAME LIQUID.*) THE REACTION GENERATED GAS THAT WOULD BUBBLE UP THROUGH THE LIQUID AND COLLECT IN THE BOTTLE.

PRIESTLEY STORED THE GASES IN PIG BLADDERS HE HAPPENED TO HAVE LYING AROUND THE HOUSE.

I'VE INVENTED THE WHOOPIE CUSHION!

*WATER, UNLESS THE GAS WAS WATER SOLUBLE, IN WHICH CASE PRIESTLEY USED MERCURY.

FOR EXAMPLE, WHEN HE COMBINED A STRONG ACID WITH IRON FILINGS, THE REACTION PRODUCED A GAS, OR "INFLAMMABLE AIR," THAT BURNED EXPLOSIVELY. WE KNOW IT AS **HYDROGEN.**

ANOTHER EXPERIMENT HEATED A RED MINERAL CALLED "CALX OF MERCURY." AS THE "CALX" MELTED, DROPLETS OF PURE MERCURY CONDENSED ON THE WALLS OF THE VESSEL, WHILE GAS ACCUMULATED IN THE WATER BOTTLE.

(PRIESTLEY HEATED WITH LENSES TO AVOID SMOKY, ASHY FIRES.)

PRIESTLEY NOTICED THAT A FLAME BURNED EXTRA BRIGHTLY WHEN SURROUNDED BY THIS NEW GAS.

SINCE HE KNEW THAT FLAMES BURN WELL IN "GOOD" (I.E., BREATHABLE) AIR AND SNUFF OUT IN BAD AIR (AS IN A COAL MINE), HE DECIDED TO TAKE A WHIFF.

AFTERWARD, HE WROTE:

"THE FEELING OF IT TO MY LUNGS WAS NOT SENSIBLY DIFFERENT FROM THAT OF COMMON AIR. BUT I FANCIED THAT MY BREATH FELT PARTICULARLY LIGHT AND EASY FOR SOME TIME AFTERWARD. WHO CAN TELL BUT THAT, IN TIME, THIS PURE AIR MAY BECOME A FASHIONABLE ARTICLE IN LUXURY? HITHERTO ONLY TWO MICE AND MYSELF HAVE HAD THE PRIVILEGE OF BREATHING IT."

NOW I'VE INVENTED THE OXYGEN BAR!

FOR OXYGEN IT WAS...

AT THE SAME TIME, IN FRANCE, **ANTOINE LAVOISIER** (1743 – 1794) WAS DOING A SIMILAR EXPERIMENT, BUT IN REVERSE.

LAVOISIER HEATED A PIECE OF METALLIC TIN IN A TIGHTLY SEALED FLASK. A GRAYISH ASH APPEARED ON THE SURFACE OF THE MELTING TIN. LAVOISIER HEATED IT FOR A DAY AND A HALF UNTIL NO MORE ASH FORMED.

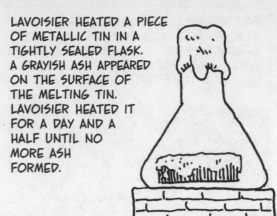

AFTER ALLOWING THE FLASK TO COOL, HE INVERTED IT AND UNSEALED IT UNDER WATER.

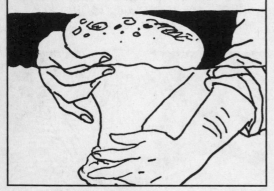

HE NOTED THAT THE WATER ROSE **ONE-FIFTH OF THE WAY** INTO THE FLASK.

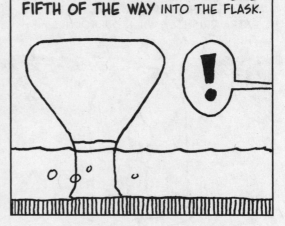

CONCLUSION: ONE-FIFTH OF THE AIR ORIGINALLY IN THE FLASK WAS REMOVED BY THE REACTION. THIS GAS MUST HAVE COMBINED WITH THE TIN TO FORM THE ASHY SUBSTANCE.

AIR, SAID LAVOISIER, MUST BE A **MIXTURE** OF TWO DIFFERENT GASES. ONE OF THEM, WHICH MAKES UP ONE-FIFTH OF THE TOTAL VOLUME, COMBINED WITH THE TIN, WHILE THE OTHER DID NOT.

IN OTHER WORDS, AIR IS NOT AN ELEMENT!

D'OH!

NEXT LAVOISIER REPEATED THE EXPERIMENT USING MERCURY INSTEAD OF TIN. OVER HIGH HEAT, MERCURY ALSO FORMED AN ASH (OR "CALX") AND REMOVED GAS FROM THE AIR. THEN, WHEN HEATED GENTLY, THE ASH GAVE BACK THE GAS AND ALL THE ORIGINAL MERCURY, A LA PRIESTLEY.

THE EXPERIMENT IS REVERSIBLE!

IN OTHER WORDS, PRIESTLEY'S "GOOD AIR" WAS THE SAME GAS THAT LAVOISIER HAD FOUND TO MAKE UP 20% OF THE ATMOSPHERE. THE FRENCH CHEMIST GAVE IT A NEW NAME: **OXYGEN.**

IT'S EVERY-WHERE!

INTERPRETATION: THE ASH WAS A **COMPOUND** OF THE METAL AND OXYGEN (A METALLIC OXIDE, WE WOULD SAY).

THE OXYGEN COMES FROM THE AIR IN THE FLASK.

LAVOISIER CONFIRMED THIS BY WEIGHING: THE WEIGHT OF THE REMAINING (UNREACTED) METAL PLUS THE WEIGHT OF ASH WAS **GREATER** THAN THE WEIGHT OF THE ORIGINAL METAL.

THE EXTRA WEIGHT COMES FROM THE OXYGEN!

LAVOISIER DREW A GENERAL CONCLUSION: **COMBUSTION** WAS A PROCESS WHEREBY FUEL COMBINED WITH OXYGEN. IN OTHER WORDS, **FIRE IS NOT AN ELEMENT;** IT'S A CHEMICAL REACTION THAT GOBBLES UP OXYGEN AND GIVES OFF HEAT AND LIGHT.

WHAT? NOW WE'RE DOWN TO **TWO** ELEMENTS?

SORRY, ARI....

AND MORE: LAVOISIER ALSO FOUND THAT THE TOTAL WEIGHT OF THE SEALED FLASK PLUS CONTENTS WAS THE SAME BEFORE AND AFTER THE REACTION.

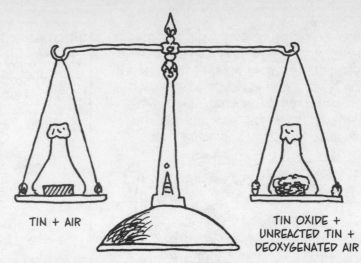

TIN + AIR

TIN OXIDE + UNREACTED TIN + DEOXYGENATED AIR

AND SO HE LAID DOWN THE LAW OF **CONSERVATION OF MATTER.**

In chemical reactions, nothing is created or destroyed. The elements are merely rearranged in new combinations.

LAVOISIER PROPOSED A PROGRAM FOR CHEMISTRY: FIND THE ELEMENTS, THEIR WEIGHTS, AND THEIR RULES OF COMBINATION. THEN HE LOST HIS HEAD IN THE FRENCH REVOLUTION, AND THE PROGRAM, LIKE HIS HEAD, HAD TO BE CARRIED OUT BY OTHERS.

THE WEIGHT OF THE HEAD PLUS THE WEIGHT OF THE BODY...

CHEMISTS FOLLOWED THROUGH WITH ENTHUSIASM, AND BY 1800 HAD DISCOVERED ABOUT THIRTY ELEMENTS—AND NONE OF THEM WAS **WATER.** IT TURNED OUT TO BE A COMPOUND OF HYDROGEN AND OXYGEN.

CARE FOR A HYDROGEN BALLOON? FUN!

GRRRR...

AND ONE MORE WAY YOU'RE WRONG...

SIGH...

AND **COMPOUNDS**, THEY FOUND, WERE NO MERE ARISTOTELIAN MISH-MASH. INSTEAD, COMPOUNDS ALWAYS COMBINED ELEMENTS IN **FIXED PROPORTIONS.** WATER, FOR EXAMPLE, WAS ALWAYS MADE OF EXACTLY **TWO** VOLUMES OF HYDROGEN GAS AND **ONE** VOLUME OF OXYGEN.

AS A COOK, NATURE IS OBSESSIVE-COMPULSIVE!

SUCH DISCOVERIES LED **JOHN DALTON** (1766–1844) TO REVIVE THE **ATOMIC THEORY OF MATTER.** EACH ELEMENT, HE REASONED, WAS MADE OF TINY, INDIVISIBLE ATOMS. THE ATOMS OF ANY ONE ELEMENT ARE ALL ALIKE, BUT DIFFER FROM THE ATOMS OF OTHER ELEMENTS.

COMPOUND SUBSTANCES, SAID DALTON, WERE COMPOSED OF FIXED GROUPINGS OF ATOMS CALLED **MOLECULES.**

ALTHOUGH ATOMS WERE INVISIBLY SMALL, SCIENTISTS ACCEPTED THE ATOMIC THEORY ANYWAY, BECAUSE IT EXPLAINED WHAT THEY COULD SEE...

13

MEANWHILE, THEY KEPT UP THE HUNT FOR NEW ELEMENTS, FINDING NEARLY **SEVENTY** BY THE 1860s—AND WHAT A LIST IT WAS! ELEMENTS MIGHT BE SOLID, LIQUID, OR GASEOUS; YELLOW, GREEN, BLACK, WHITE, OR COLORLESS; CRUMBLY OR BENDY; WILDLY REACTIVE OR RELATIVELY INERT.

WHERE'S THE SENSE IN IT?

ONE THING SOON BECAME CLEAR: SOME ELEMENTS WERE MORE ALIKE THAN OTHERS. **SODIUM** AND **POTASSIUM** BOTH REACTED VIOLENTLY WITH WATER. **CHLORINE, FLUORINE,** AND **BROMINE** ALL COMBINED ON A ONE-TO-ONE BASIS WITH SODIUM AND POTASSIUM. **CARBON** AND **SILICON** BOTH HOOKED UP WITH TWO **OXYGENS**... ETC.

HM! ELEMENTS HAVE **FAMILIES,** JUST LIKE PEOPLE!

SODIUM REMINDS ME OF MY MOM: BITTER AND REACTIVE...

ONE MORNING IN 1869, A RUSSIAN NAMED **DMITRI MENDELEEV** (1834–1907) WOKE UP WITH AN IDEA: LIST THE ELEMENTS IN ORDER OF INCREASING ATOMIC WEIGHT AND DO A "TEXT WRAP" AT REGULAR INTERVALS.

HAVEN'T YOU EVER HAD THAT DREAM?

THE RESULT WAS A SORT OF TABLE, WITH THE ELEMENTS ARRANGED IN ROWS. HERE'S A BABY VERSION OF MENDELEEV'S TABLE. (YOU'LL SEE THE REAL THING NEXT CHAPTER.)

HYDROGEN						
LITHIUM	BERYLLIUM	BORON	CARBON	NITROGEN	OXYGEN	FLUORINE
SODIUM	MAGNESIUM	ALUMINUM	SILICON	PHOSPHORUS	SULFUR	CHLORINE
POTASSIUM	CALCIUM					

THE ELEMENTS SHOWED A **PERIODIC PATTERN:** EACH VERTICAL COLUMN CONTAINED CHEMICALLY SIMILAR ELEMENTS. IN FACT, MENDELEEV NOTED GAPS FARTHER DOWN THE TABLE AND SUCCESSFULLY PREDICTED **NEW ELEMENTS** THAT WOULD FILL THEM!

FINE. NOW WHERE'S MY TEDDY BEAR?

THE TABLE WAS GREAT, BUT HOW TO EXPLAIN IT? IN FACT, HOW TO EXPLAIN ANY OF CHEMISTRY? WHAT ACCOUNTED FOR ATOMIC WEIGHTS, OR WHICH ELEMENTS COMBINED WITH WHICH? CHEMISTS HAD COME FAR IN INTERPRETING THEIR OBSERVATIONS, BUT A QUESTION STILL HUNG IN THE AIR: **WHY?**

LOVE THAT QUESTION!

TO FIND THE ANSWER, SCIENTISTS FOLLOWED THE SAME LINE OF THOUGHT THEY'D BEEN USING ALL ALONG: IF SUBSTANCES ARE MADE OF ELEMENTS, AND ELEMENTS ARE MADE OUT OF ATOMS, THEN WHAT, THEY ASKED, ARE ATOMS MADE OUT OF?

I DON'T KNOW WHAT ELSE TO ASK!

Chapter 2
Matter Becomes Electric

NATURE HAD ANOTHER SECRET
BESIDES FIRE... AT LEAST, IT
LOOKED LIKE ANOTHER
SECRET AT FIRST...

I'VE GOT A
MILLION OF 'EM...

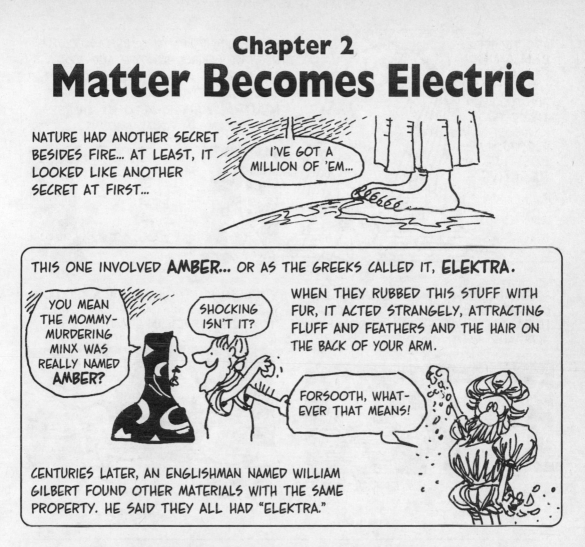

THIS ONE INVOLVED **AMBER**... OR AS THE GREEKS CALLED IT, **ELEKTRA**.

YOU MEAN
THE MOMMY-
MURDERING
MINX WAS
REALLY NAMED
AMBER?

SHOCKING
ISN'T IT?

WHEN THEY RUBBED THIS STUFF WITH
FUR, IT ACTED STRANGELY, ATTRACTING
FLUFF AND FEATHERS AND THE HAIR ON
THE BACK OF YOUR ARM.

FORSOOTH, WHAT-
EVER THAT MEANS!

CENTURIES LATER, AN ENGLISHMAN NAMED WILLIAM
GILBERT FOUND OTHER MATERIALS WITH THE SAME
PROPERTY. HE SAID THEY ALL HAD "ELEKTRA."

THEN PEOPLE NOTICED THAT THERE WERE REALLY TWO KINDS OF "ELECTRIC"
MATERIALS: ONE REPELLED WHAT THE OTHER ATTRACTED, AND VICE VERSA.

SORRY.

AROUND 1750, **BENJAMIN FRANKLIN** (1706–1790) FIRST CALLED THESE TWO KINDS OF ELECTRICITY **POSITIVE** AND **NEGATIVE.**

POSITIVE, SAID FRANKLIN, REPELS POSITIVE; NEGATIVE REPELS NEGATIVE; AND POSITIVE AND NEGATIVE ATTRACT EACH OTHER AND CANCEL EACH OTHER OUT. IN ORDINARY, **NEUTRAL** MATTER, OPPOSITE CHARGES ARE PRESENT IN EQUAL AMOUNT.

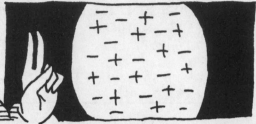

NEGATIVE CHARGES CAN SOMETIMES FLOW OUT OF A SUBSTANCE, CREATING A CHARGE **IMBALANCE**—AN EXCESS OF NEGATIVITY HERE AND POSITIVITY THERE...

BUT BECAUSE OF THE MUTUAL ATTRACTION, THE NEGATIVES MAY SUDDENLY FLOW BACK TO THE POSITIVE CHARGE WITH A SPARK.

"TWO NIGHTS AGO, BEING ABOUT TO KILL A TURKEY BY THE SHOCK FROM TWO LARGE GLASS JARS,* CONTAINING AS MUCH ELECTRICAL FIRE AS FORTY COMMON PHIALS, I INADVERTENTLY TOOK THE WHOLE THROUGH MY OWN ARMS AND BODY, BY RECEIVING THE FIRE FROM THE UNITED TOP WIRES WITH ONE HAND WHILE THE OTHER HELD A CHAIN CONNECTED WITH THE OUTSIDE OF BOTH JARS."

—BENJAMIN FRANKLIN, 1750

AND NOW FOR SOMETHING REALLY BIG!

*JUST ONE OF THE WAYS THE FUN-LOVING FOUNDING FATHER LIKED TO AMUSE HIMSELF!

WITH THE INVENTION OF THE ELECTRIC BATTERY (BY VOLTA IN 1800), ONE COULD RUN A STEADY STREAM OF NEGATIVE CHARGE—A **CURRENT**—THROUGH A COPPER WIRE, AND MAYBE THROUGH OTHER MATERIALS AS WELL.

CHEMISTS TRIED RUNNING ELECTRICITY THROUGH ORDINARY WATER. TWO METAL STRIPS, OR ELECTRODES, WERE CONNECTED TO A BATTERY AND IMMERSED IN WATER.

AS CHARGE BUILT UP ON THE ELECTRODES, BUBBLES OF **HYDROGEN GAS** APPEARED AT THE NEGATIVE STRIP, OR **CATHODE**. BUBBLES OF **OXYGEN** FORMED AT THE POSITIVE STRIP, OR ANODE.

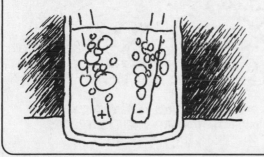

ELECTRICITY SPLITS WATER! SCIENTISTS SOON TRIED THIS **ELECTROLYSIS** (ELECTRIC SPLITTING) ON OTHER SUBSTANCES. MELTED TABLE SALT, THEY FOUND, YIELDS METALLIC **SODIUM** AT THE CATHODE AND GREEN, TOXIC **CHLORINE GAS** AT THE ANODE.

CHOKE!

IT'S A BIG LEAP FROM FINDING ELECTRICITY IN A FEW PLACES TO SEEING IT EVERYWHERE, BUT THAT'S SCIENCE FOR YOU!

LONG LIVE THE INDUCTIVE METHOD!

OBSERVATION

CALUMNY

FAILURE

RIDICULE

OBSCURITY

HYPOTHESIS

BY THE END OF THE 19TH CENTURY, SCIENTISTS WERE CONVINCED THAT ATOMS WERE MADE OF **ELECTRIC INGREDIENTS**.

AND SO THEY ARE. HERE'S THE IDEA:

ATOMS ARE MADE UP OF SMALLER, ELECTRICALLY CHARGED PARTICLES (AND SOME NEUTRAL PARTICLES TOO). EACH ATOM HAS AN EQUAL NUMBER OF POSITIVE AND NEGATIVE CHARGES. THE NEGATIVELY CHARGED PARTICLES, CALLED **ELECTRONS,** WEIGH LITTLE AND MOVE AROUND EASILY.

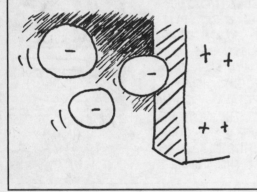

OTHER KINDS OF ATOMS **ACQUIRE** ELECTRONS TO BECOME NEGATIVELY CHARGED IONS, OR **ANIONS,** ATTRACTED TO ANODES.

A DEPARTING ELECTRON LEAVES BEHIND A POSITIVELY CHARGED ATOM, OR **POSITIVE ION.** SUCH IONS, ATTRACTED TO CATHODES (WHICH ARE NEGATIVE), ARE CALLED **CATIONS** (PRONOUNCED "CAT-EYE-ONZ").

IN TABLE SALT, FOR EXAMPLE, SODIUM CATIONS ARE ATTRACTED TO CHLORIDE ANIONS AND ARRANGE THEMSELVES INTO A CRYSTAL, **SODIUM CHLORIDE.**

DURING ELECTROLYSIS, THESE IONS MIGRATE TOWARD THE ELECTRODES, AND THE SALT DISSOCIATES.

All-Important Fact:

 ATOMS COMBINE CHEMICALLY BY SHARING OR TRANSFERRING ELECTRONS.

SO—TO UNDERSTAND CHEMISTRY, WE NEED TO SEE HOW ELECTRONS BEHAVE WITHIN EACH ATOM.

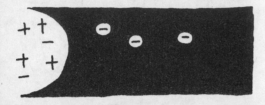

HOW SMALL IS THE SMALL PICTURE? LET'S TRY SHRINKING DOWN **ONE MILLION TIMES.** A HUMAN HAIR IS NOW THIRTY STORIES THICK... BACTERIA ARE THE SIZE OF TORPEDOES... AND ATOMS ARE JUST BARELY VISIBLE AS TINY SPECKS.

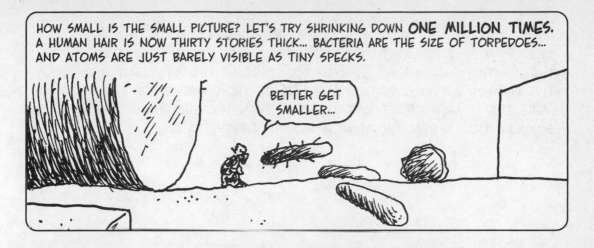

BETTER GET SMALLER...

SHRINKING ANOTHER THOUSAND TIMES BRINGS US TO **NANOMETER** (= 10^{-9} METER) SCALE. I'M JUST SHY OF 2 nm TALL. THE ATOMS ARE NOW ABOUT ONE-TENTH MY SIZE. WE'RE IN A VERY ENERGETIC ENVIRONMENT: LIGHT WAVES ARE ZOOMING AROUND, AND ALL THE ATOMS ARE JIGGLING.

THIS IS **GRAPHITE** FROM SOME PENCIL SHAVINGS. THE CARBON ATOMS ARE ARRANGED IN SHEETS THAT CAN SLIDE OVER EACH OTHER EASILY. THIS EXPLAINS WHY GRAPHITE IS A GOOD LUBRICANT.*

BUT STILL NOTHING ELECTRIC!

LET'S SHRINK TEN MORE TIMES TO ATOMIC SIZE—10^{-10} METER—AND LOOK AT A SINGLE CARBON ATOM. I CAN VAGUELY SENSE SOME ELECTRONS HUMMING AROUND, ALTHOUGH THEY'RE AWFULLY HARD TO PIN DOWN.

BUT WHERE ARE THE POSITIVE CHARGES?

*IN PURE FORM. PENCIL LEAD IS A MIXTURE OF GRAPHITE AND CLAY.

NOW I'M A HUNDRED TIMES SMALLER, AT **PICOMETER** SCALE. THAT'S A MILLIONTH OF A MILLIONTH, OR 10^{-12} ACTUAL SIZE. THERE AT LAST ARE THE POSITIVE CHARGES, ALL LUMPED TOGETHER AT THE VERY CENTER OF THE ATOM IN A TINY CORE OR **NUCLEUS.** IF THE DIAMETER OF THE ATOM WERE THE LENGTH OF A FOOTBALL FIELD, THEN THE NUCLEUS WOULD BE SMALLER THAN A PEA. THE ATOM IS MOSTLY EMPTY SPACE!

10^{-12} m ⟶

ORDINARILY, THE CARBON NUCLEUS CONSISTS OF TWELVE PARTICLES: SIX **PROTONS** WITH A POSITIVE CHARGE AND SIX **NEUTRONS** WITH NO CHARGE AT ALL. THE PROTONS' CHARGE IS BALANCED BY THE SIX HOVERING NEGATIVE ELECTRONS, SO THE ATOM IS NEUTRAL OVERALL.

BUT WHY DON'T THE PROTONS REPEL EACH OTHER?

THE NUCLEUS IS HELD TOGETHER BY A POWERFUL, SHORT-RANGE ATTRACTION CALLED THE **STRONG FORCE,** * WHICH OVERCOMES ELECTRICAL REPULSION. THIS INTENSE PULL MAKES MOST NUCLEI VIRTUALLY INDESTRUCTIBLE. THIS VERY SAME CARBON ATOM HAS BEEN ROAMING THE EARTH FOR BILLIONS OF YEARS.

DINOSAUR

POOP

AIR

FLY

YOUR NOSE

PLANT

GROUND

NEARLY ALL THE ATOM'S MASS IS CONCENTRATED IN THE TINY NUCLEUS. EACH PROTON AND NEUTRON (THEY HAVE ALMOST EXACTLY THE SAME WEIGHT) HAS 1840 TIMES THE MASS OF AN ELECTRON.

TO BE PRECISE:

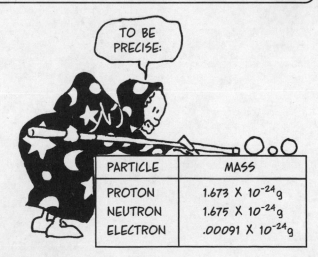

PARTICLE	MASS
PROTON	1.673×10^{-24} g
NEUTRON	1.675×10^{-24} g
ELECTRON	$.00091 \times 10^{-24}$ g

*SCIENTISTS DON'T INVENT NEARLY SUCH COLORFUL NAMES AS THEY USED TO.

NOW FOR A FEW HELPFUL

definitions:

AN ELEMENT'S **ATOMIC NUMBER** IS THE NUMBER OF PROTONS IN ITS NUCLEUS. CARBON'S ATOMIC NUMBER IS 6.

AMAZING! HOW DO YOU KNOW?

MASS SPECTROSCOPY... CHEMICAL ANALYSIS... CAREFUL WEIGHING... AND I COUNTED!

ALMOST 99% OF ALL CARBON ATOMS ON EARTH HAVE SIX NEUTRONS ALONG WITH THEIR SIX PROTONS. WE CALL THIS CARBON-12 (AND SOMETIMES WRITE ^{12}C), SINCE ITS MASS IS SO CLOSE TO THAT OF TWELVE NUCLEAR PARTICLES.

MORE PRECISELY, CHEMISTS DEFINE AN **ATOMIC MASS UNIT,** OR **AMU,** TO BE PRECISELY **ONE-TWELFTH THE MASS OF A** ^{12}C **ATOM.** THE COMMON CARBON ATOM HAS A MASS OF EXACTLY 12.000000 AMU, BY DEFINITION. ALL OTHER ATOMIC MASSES ARE COMPUTED RELATIVE TO THIS REFERENCE.

EACH PROTON AND NEUTRON WEIGHS ABOUT ONE AMU.

THE OTHER 1.1% OF CARBON ATOMS HAVE SEVEN NEUTRONS. THERE MUST STILL BE SIX PROTONS (OTHERWISE IT'S NOT CARBON!), BUT THIS **CARBON-13** ATOM WEIGHS APPRECIABLY MORE THAN CARBON-12.

^{12}C, ^{13}C, AND A VERY RARE FORM, ^{14}C, WITH EIGHT NEUTRONS, ARE CALLED **ISOTOPES** OF CARBON. THE ISOTOPES OF AN ELEMENT HAVE THE SAME NUMBER OF PROTONS, BUT DIFFERENT NUMBERS OF NEUTRONS.

^{13}C NUCLEUS

^{14}C NUCLEUS

THE SIMPLEST ATOM OF ALL IS **HYDROGEN,** SYMBOL H, WITH AN ATOMIC NUMBER OF ONE. IN NEARLY ALL HYDROGEN ATOMS, A SINGLE ELECTRON ORBITS A SINGLE PROTON, BUT ISOTOPES WITH ONE AND TWO NEUTRONS ALSO EXIST.

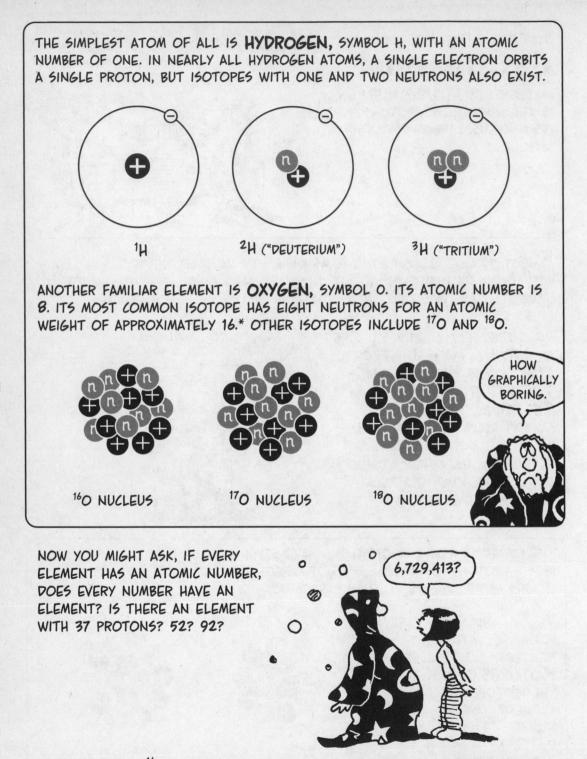

^{1}H ^{2}H ("DEUTERIUM") ^{3}H ("TRITIUM")

ANOTHER FAMILIAR ELEMENT IS **OXYGEN,** SYMBOL O. ITS ATOMIC NUMBER IS 8. ITS MOST COMMON ISOTOPE HAS EIGHT NEUTRONS FOR AN ATOMIC WEIGHT OF APPROXIMATELY 16.* OTHER ISOTOPES INCLUDE ^{17}O AND ^{18}O.

HOW GRAPHICALLY BORING.

^{16}O NUCLEUS ^{17}O NUCLEUS ^{18}O NUCLEUS

NOW YOU MIGHT ASK, IF EVERY ELEMENT HAS AN ATOMIC NUMBER, DOES EVERY NUMBER HAVE AN ELEMENT? IS THERE AN ELEMENT WITH 37 PROTONS? 52? 92?

6,729,413?

*THE ACTUAL MASS OF ^{16}O IS 15.9949 AMU. THE "MISSING MASS" IS CONVERTED TO THE **ENERGY** OF THE STRONG FORCE THAT BINDS THE NUCLEUS TOGETHER. OTHER ATOMS HAVE SIMILAR FRACTIONAL WEIGHTS.

NATURE, IT TURNS OUT, MAKES ATOMS WITH EVERY NUMBER FROM 1 (HYDROGEN) TO 92 (URANIUM), ALTHOUGH A FEW ELEMENTS IN THERE ARE VERY SCARCE.

THE SEQUENCE STOPS THERE BECAUSE LARGE NUCLEI (THOSE ABOVE 83, BISMUTH) ARE UNSTABLE. BEYOND URANIUM, 92, THEY FALL APART SO QUICKLY THAT WE DON'T SEE THEM IN NATURE. PHYSICISTS CAN MAKE NUCLEI WITH MORE THAN 92 PROTONS, BUT THEY DON'T SURVIVE LONG.

HERE IS A LIST OF THE 92 NATURALLY OCCURRING ELEMENTS:

1. Hydrogen, H
2. Helium, He
3. Lithium, Li
4. Beryllium, Be
5. Boron, B
6. Carbon, C
7. Nitrogen, N
8. Oxygen, O
9. Fluorine, F
10. Neon, Ne
11. Sodium, Na
12. Magnesium, Mg
13. Aluminum, Al
14. Silicon, Si
15. Phosphorus, P
16. Sulfur, S
17. Chlorine, Cl
18. Argon, Ar
19. Potassium, K
20. Calcium, Ca
21. Scandium, Sc
22. Titanium, Ti
23. Vanadium, V
24. Chromium, Cr
25. Manganese, Mn
26. Iron, Fe
27. Cobalt, Co
28. Nickel, Ni

29. Copper, Cu
30. Zinc, Zn
31. Gallium, Ga
32. Germanium, Ge
33. Arsenic, As
34. Selenium, Se
35. Bromine, Br
36. Krypton, Kr
37. Rubidium, Rb
38. Strontium, Sr
39. Yttrium, Y
40. Zirconium, Zr
41. Niobium, Nb
42. Molybdenum, Mo
43. Technetium, Tc
44. Ruthenium, Ru
45. Rhodium, Rh
46. Palladium, Pd
47. Silver, Ag
48. Cadmium, Cd
49. Indium, In
50. Tin, Sn
51. Antimony, Sb
52. Tellurium, Te
53. Iodine, I
54. Xenon, Xe
55. Cesium, Cs
56. Barium, Ba

57. Lanthanum, La
58–71—Never mind these!
72. Hafnium, Hf
73. Tantalum, Ta
74. Tungsten, W
75. Rhenium, Re
76. Osmium, Os
77. Iridium, Ir
78. Platinum, Pt
79. Gold, Au
80. Mercury, Hg
81. Thallium, Tl
82. Lead, Pb
83. Bismuth, Bi
84. Polonium, Po
85. Astatine, At
86. Radon, Rn
87. Francium, Fr
88. Radium, Ra
89. Actinium, Ac
90. Thorium, Th
91. Protactinium, Pa
92. Uranium, U

(93, 94, AND ABOVE ARE ARTIFICIAL AND UNSTABLE.)

The Elusive Electron

TO TURN THAT RATHER STARK LIST INTO A PERIODIC TABLE—FOR THAT IS OUR GOAL—WE NOW TURN TO THE ATOM'S OTHER MAIN INGREDIENT, ITS ELECTRONS. THESE, WE SHOULD WARN YOU, DEFY COMMON SENSE, BECAUSE ELECTRONS, YOU SEE, OBEY THE BIZARRE RULES OF MODERN PHYSICS CALLED **QUANTUM MECHANICS.**

WRAP YOUR MIND AROUND THIS: AN ELECTRON IS A **PARTICLE**, LIKE A MARBLE, BUT ALSO A **WAVE,** LIKE A BEAM OF LIGHT. AS A PARTICLE, IT HAS A DEFINITE **MASS, CHARGE,** AND **SPIN,** BUT IT ALSO HAS A **WAVELENGTH.** IT'S "SMEARED OUT" IN SOME WAY. ITS PRECISE POSITION IS ALWAYS A BIT UNCERTAIN. MAKE SENSE? WE DIDN'T THINK SO!

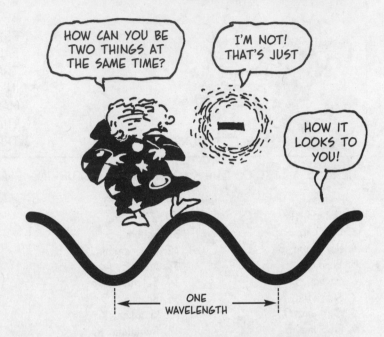

HOW CAN YOU BE TWO THINGS AT THE SAME TIME?

I'M NOT! THAT'S JUST

HOW IT LOOKS TO YOU!

ONE WAVELENGTH

IN ITS GUISE AS A PARTICLE, AN ELECTRON INHABITS A SORT OF "PROBABILITY CLOUD"—**NOT** A CIRCULAR ORBIT. THE DENSEST PARTS OF THE CLOUD ARE WHERE THE ELECTRON IS LIKELIEST TO "BE"—IF IT CAN BE SAID TO BE ANYWHERE, WHICH IT CAN'T EXACTLY. THESE CLOUDS NEED NOT BE ROUND, BY THE WAY.

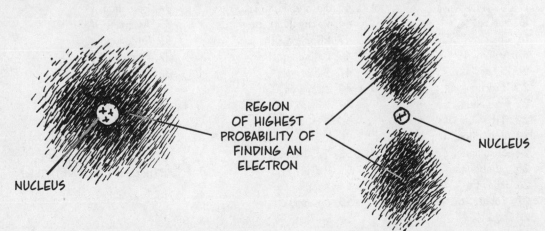

NUCLEUS

REGION OF HIGHEST PROBABILITY OF FINDING AN ELECTRON

NUCLEUS

NUCLEUS

WE CAN ALSO VISUALIZE THE ELECTRON AS A WAVE, BEAMING AROUND THE NUCLEUS. IN THIS PICTURE, QUANTUM MECHANICS TELLS US THAT THE ELECTRON IS ALWAYS A "STANDING WAVE." THAT IS, IT "GOES AROUND" THE NUCLEUS A **WHOLE NUMBER OF WAVELENGTHS**: 1, 2, 3, 4, ETC., BUT NEVER A FRACTIONAL VALUE.

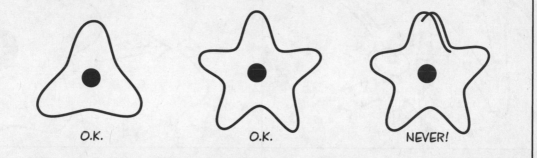

O.K.   O.K.   NEVER!

IN OTHER WORDS, ONLY CERTAIN DISCRETE "ORBITS" ARE AVAILABLE TO AN ELECTRON IN AN ATOM.

LET'S CONTRAST THIS WITH A MORE FAMILIAR SYSTEM: A PLANET ORBITING A STAR.

IMAGINE THAT SOMETHING GIVES THE PLANET A NUDGE, ADDING ENERGY TO IT.

THE EXTRA ENERGY PUSHES THE PLANET INTO AN ORBIT FARTHER FROM THE STAR.

NEW HIGHER-ENERGY ORBIT

OLD ORBIT

IN FACT, WITH A BIG ENOUGH JOLT, THE PLANET WILL ESCAPE THE STAR'S GRAVITATIONAL PULL COMPLETELY.

AN ORBITING ELECTRON IS SIMILAR: IT MAY ABSORB A JOLT OF ENERGY, TOO, IN THE FORM OF A BEAM OF LIGHT, FOR EXAMPLE.

BUT THE ELECTRON MUST JUMP TO AN ORBIT CONSISTENT WITH A WHOLE NUMBER OF WAVELENGTHS.

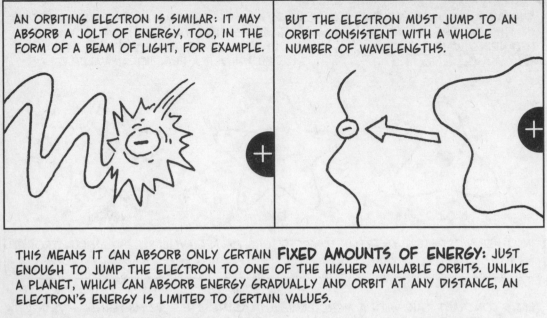

THIS MEANS IT CAN ABSORB ONLY CERTAIN **FIXED AMOUNTS OF ENERGY:** JUST ENOUGH TO JUMP THE ELECTRON TO ONE OF THE HIGHER AVAILABLE ORBITS. UNLIKE A PLANET, WHICH CAN ABSORB ENERGY GRADUALLY AND ORBIT AT ANY DISTANCE, AN ELECTRON'S ENERGY IS LIMITED TO CERTAIN VALUES.

PLANET: ALL ORBITS ARE POSSIBLE

ELECTRON: ONLY SOME ORBITS ARE POSSIBLE

WE SAY THE ELECTRON'S ENERGY IS **QUANTIZED:** IN ANY GIVEN ATOM, THE ELECTRONS CAN ASSUME ONLY CERTAIN FIXED, DISCRETE ENERGY LEVELS.

I HAVE ONLY ONE ENERGY LEVEL...

THE ELECTRON CONFIGURATIONS WITHIN EACH ENERGY LEVEL ARE CALLED **ORBI-TALS** (NAMED, NO DOUBT, BY NOSTALGIC PHYSICISTS DREAMING OF PLANETS).

SHALL WE CALL THEM... ORBITOIDS? ORBISCUITS? ORBITUARIES?

I GO AROUND AND AROUND!

THE SIMPLEST EXAMPLE IS **HYDROGEN:** ONE ELECTRON PULLED BY A SINGLE PROTON. THE ELECTRON CAN INHABIT ANY ONE OF SEVEN DIFFERENT LEVELS, OR "SHELLS," MISLEADINGLY DEPICTED HERE AS CIRCULAR ORBITS.

THIS GRAPH SHOWS THE ELECTRON'S ENERGY IN EACH SHELL.

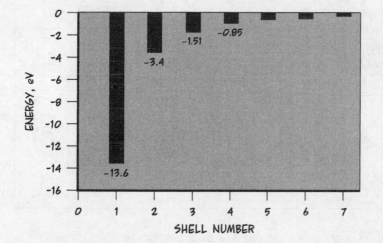

THE ENERGY UNIT HERE IS THE **ELECTRON VOLT** (eV). ONE eV IS THE ENERGY GAINED BY ONE ELECTRON PUSHED BY ONE VOLT. (NOTE: IN ATOMS, AN ELECTRON'S ENERGY IS NEGATIVE, SINCE ENERGY MUST BE ADDED TO PULL THE ELECTRON FREE OF THE NUCLEUS. THE FREE STATE IS TAKEN TO HAVE ENERGY = 0.)

TO RAISE AN ELECTRON FROM SHELL 1 TO SHELL 2 REQUIRES AN ENERGY EQUAL TO THE DIFFERENCE $(-3.4) - (-13.6) = 13.6 - 3.4 = $ **10.2** eV.

ALL THOSE MINUS SIGNS!

TO REMOVE THE ELECTRON COMPLETELY AND MAKE A HYDROGEN ION REQUIRES **13.6** eV. THIS IS CALLED THE ATOM'S **IONIZATION ENERGY.**

IONIZATION?

GESUNDHEIT.

NOW LET'S BUILD SOME BIGGER ATOMS!

LARGER ATOMS, LIKE HELIUM, LITHIUM, OR TIN, ALSO HAVE UP TO SEVEN ELECTRON SHELLS. BUT IN THESE ATOMS, THE "HIGHER" SHELLS CAN HOLD MORE ELECTRONS THAN LOWER SHELLS CAN.

THEY LOOK LIKE BALLOON DOGGIES!

HIGHER-SHELL ELECTRONS CAN ALSO HAVE MORE COMPLEX CONFIGURATIONS, OR **ORBITALS,** THAN LOWER-SHELL ELECTRONS. YOU CAN THINK OF THESE ORBITALS AS ENERGY SUBLEVELS. DIFFERENT SUBLEVELS ARE CALLED s, p, d, AND f, AND EACH ORBITAL CAN HOLD **UP TO TWO ELECTRONS.**

SHELL 1 HAS ONLY AN s ORBITAL, WHICH IS SPHERICAL. IT CAN HOLD ONE OR TWO ELECTRONS.

SHELL 2 HAS ONE s AND THREE p ORBITALS, WHICH LOOK SOMETHING LIKE DUMBBELLS. WHEN FULL, THIS SHELL HOLDS EIGHT ELECTRONS.

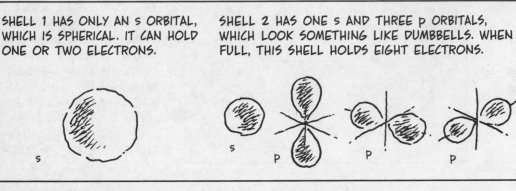

s

s   p   p   p

SHELL 3 HAS ONE s, THREE p, AND FIVE d ORBITALS (FORGET DRAWING THEM ALL!). WHEN FULL, IT HOLDS 18 ELECTRONS (2 X [1 + 3 + 5]).

s   p   p

p   d   d

AND THREE MORE d ORBITALS

SHELLS 4 AND HIGHER HAVE ALL OF THAT PLUS SEVEN f ORBITALS—UP TO 32 ELECTRONS TOTAL.

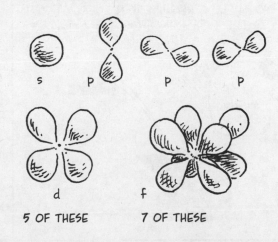

s   p   p

d   f

5 OF THESE   7 OF THESE

THE BUILDUP TO THE BUILDUP:

THIS DIAGRAM SHOWS THE ENERGY LEVELS OF THE DIFFERENT ORBITALS. THE FARTHER UP THE PAGE, THE HIGHER THE ENERGY.

NOTE THAT THE SHELLS HAVE **OVERLAPPING ENERGIES:** E.G., SOME ORBITALS IN SHELL 4 (4d AND 4f) HAVE HIGHER ENERGY THAN SOME ORBITALS IN SHELL 5 (5s), EVEN THOUGH 4 IS "LOWER" THAN 5.

NOTE: 2s MEANS THE s ORBITAL IN SHELL 2, 4d MEANS THE d ORBITAL IN SHELL 4, ETC. EACH ARROW LEADS TO THE ORBITAL WITH THE NEXT-HIGHEST ENERGY.

AS WE BUILD UP AN ATOM, EACH ELECTRON "WANTS" TO GO INTO THE LOWEST AVAILABLE ENERGY STATE. WE START AT THE LOWEST, THEN WHEN THAT FILLS UP, GO TO THE NEXT-LOWEST, ETC.

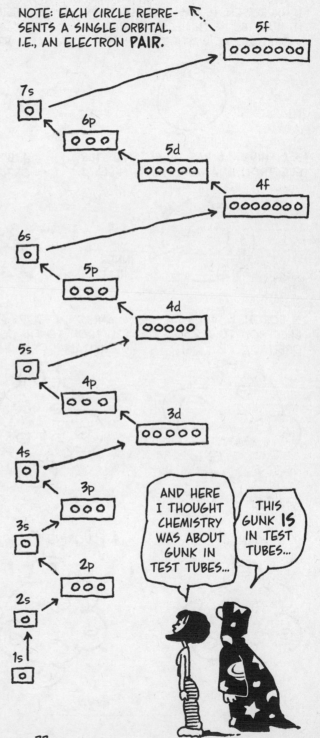

33

NOW LET'S BUILD SOME ATOMS.

1. HYDROGEN, H, HAS ONE ELECTRON. IT MUST BE IN THE LOWEST SHELL'S s ORBITAL. WE WRITE THIS AS $1s^1$.

2. HELIUM, He, ADDS A SECOND ELECTRON TO THIS s ORBITAL. NOW SHELL 1 IS FULL, AND WE WRITE $1s^2$.

$1s^1$

REMEMBER: TWO ELECTRONS PER ORBITAL, TOPS!

$1s^2$

3. LITHIUM, Li, HAS TO PUT THE THIRD ELECTRON IN A NEW SHELL, SHELL 2.

4. BERYLLIUM, Be, COMPLETES THE 2s ORBITAL.

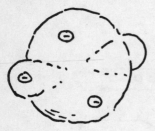

INNER SHELL

$1s^2 2s^1$

FROM HERE ON, WE OMIT THE INNER SHELL IN THE DRAWING.

$1s^2 2s^2$

5. BORON, B, ADDS AN ELECTRON TO A 2p ORBITAL.

6. CARBON, C, ADDS AN ELECTRON TO THE SECOND p ORBITAL.

7. NITROGEN, N, ADDS AN ELECTRON TO THE THIRD p ORBITAL.

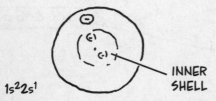

$1s^2 2s^2 2p^1$

$1s^2 2s^2 2p^2$

$1s^2 2s^2 2p^3$

8. OXYGEN, O

9. FLUORINE, F

10. NEON, Ne, COMPLETES SHELL 2.

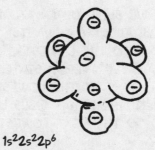

$1s^2 2s^2 2p^4$

$1s^2 2s^2 2p^5$

$1s^2 2s^2 2p^6$

TO FIND OUT WHAT HAPPENS IN ELEMENT #11, LOOK AT THE CHART ON p 33. AFTER 2p FILLS UP, THE LOWEST-ENERGY AVAILABLE ORBITAL IS 3s, IN THE THIRD SHELL, FOLLOWED BY 3p. SO WE HAVE:

11. SODIUM, Na. WE CAN WRITE THIS AS Ne3s^1, INDICATING ONE s ELECTRON ORBITING OUTSIDE A GROUP OF ELECTRONS JUST LIKE NEON'S.

12. MAGNESIUM, Mg. SIMILARLY, WE CAN WRITE THIS AS Ne3s^2.

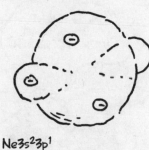

Ne3s^1

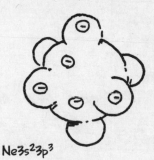

Ne3s^2

13. ALUMINUM, Al

14. SILICON, Si

15. PHOSPHORUS, P

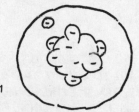

Ne3s^{2}3p^1

Ne3s^{2}3p^2

Ne3s^{2}3p^3

16. SULFUR, S

17. CHLORINE, Cl

18. ARGON, Ar

Ne3s^{2}3p^4

Ne3s^{2}3p^5

Ne3s^{2}3p^6

IF YOU COMPARE THESE ATOMS WITH THOSE ON THE PREVIOUS PAGE, YOU WILL SEE THAT ELEMENTS 11-18 ARE LIKE "BIG SISTERS" TO ELEMENTS 3-10. EACH OF THE ATOMS ON THIS PAGE HAS AN **OUTER SHELL** IDENTICAL TO THAT OF THE ATOM JUST EIGHT ELEMENTS BEHIND IT!

WE WRITE THE FIRST EIGHTEEN ELEMENTS IN A TABLE. IN ANY COLUMN, ALL THE ATOMS HAVE THE SAME OUTER ELECTRON CONFIGURATION.

1 H							2 He
3 Li	4 Be	5 B	6 C	7 N	8 O	9 F	10 Ne
11 Na	12 Mg	13 Al	14 Si	15 P	16 S	17 Cl	18 Ar

(EXCEPT HELIUM, WHICH GOES IN THE LAST COLUMN BECAUSE ITS OUTER SHELL IS FULL.)

NEXT, ACCORDING TO THE CHART ON P. 33 THE 4s ORBITAL FILLS AS WE BEGIN THE FOURTH ROW OF THE TABLE. NEXT, SAYS THE CHART, ELECTRONS BEGIN TO OCCUPY THE 3d ORBITALS. BEFORE WE CAN CONTINUE IN THE FOURTH SHELL, TEN ELECTRONS MUST GO INTO THESE **INNER** ORBITALS. WE WRITE THESE TEN ELEMENTS ON A LOOP, SINCE WE'RE STALLED FILLING THE FOURTH SHELL.

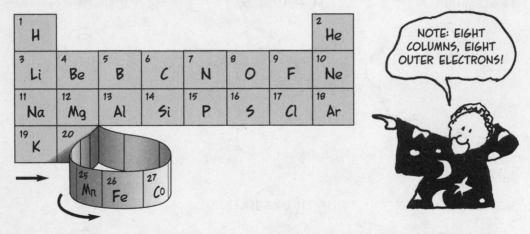

NOTE: EIGHT COLUMNS, EIGHT OUTER ELECTRONS!

AFTER THOSE TEN, WE CAN RESUME PUTTING ELECTRONS IN THE FOURTH SHELL, UNTIL ALL THE 4s AND 4p ORBITALS ARE FULL AT ELEMENT 36, KRYPTON, Kr.

AGAIN, WITHIN EACH COLUMN THAT LIES "FLAT ON THE PAGE," ATOMS HAVE OUTER SHELLS THAT LOOK THE SAME.

THE FIFTH ROW FILLS UP IN EXACTLY THE SAME WAY AS THE FOURTH: FIRST THE OUTER s, THEN THE INNER d, THEN THE OUTER p.

THE ELEMENTS THAT ARE "FLAT ON THE PAGE" ARE CALLED **MAIN-GROUP ELEMENTS.** THOSE IN THE LOOPS ARE CALLED **TRANSITION METALS.**

THE SIXTH ROW HAS A LOOP WITHIN A LOOP, AS 4f ORBITALS FILL BEFORE 5d. (SEE P. 33!) AS THERE ARE SEVEN 4f ORBITALS, THIS LOOP HAS 14 ELEMENTS. IT IS CALLED THE **LANTHANIDE SERIES,** AFTER ITS FIRST ELEMENT, LANTHANUM.

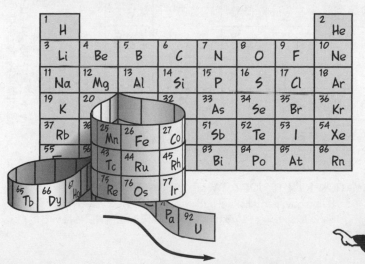

THE SEVENTH ROW PETERS OUT WHEN WE RUN OUT OF ELEMENTS.

AND THAT IS THE END OF OUR TABLE!

TURN THIS PAGE SIDEWAYS TO SEE THE PERIODIC TABLE AS IT IS USUALLY DISPLAYED. THE d-LOOPS ARE FLATTENED OUT TO SHOW EVERY ELEMENT. THE 14-ELEMENT f-LOOP, AFTER 57, LANTHANUM, IS CUT OUT AND PUT BELOW THE MAIN TABLE. THE TABLE'S "TAIL," THE ACTINIDE SERIES AFTER 89, IS ALSO AT THE BOTTOM.

1 H 1.01																	2 He 4.00
3 Li 6.94	4 Be 9.01											5 B 10.81	6 C 12.01	7 N 14.01	8 O 16.00	9 F 19.00	10 Ne 20.18
11 Na 22.99	12 Mg 24.31											13 Al 26.98	14 Si 28.09	15 P 30.97	16 S 32.07	17 Cl 35.45	18 Ar 39.95
19 K 39.10	20 Ca 40.08	21 Sc 44.96	22 Ti 47.88	23 V 50.94	24 Cr 52.00	25 Mn 54.94	26 Fe 55.85	27 Co 58.93	28 Ni 58.69	29 Cu 63.55	30 Zn 65.39	31 Ga 69.72	32 Ge 72.59	33 As 74.92	34 Se 78.96	35 Br 79.90	36 Kr 83.80
37 Rb 85.47	38 Sr 87.62	39 Y 88.91	40 Zr 91.22	41 Nb 92.91	42 Mo 95.94	43 Tc (98)	44 Ru 101.1	45 Rh 102.9	46 Pd 106.4	47 Ag 107.9	48 Cd 112.4	49 In 114.8	50 Sn 118.7	51 Sb 121.8	52 Te 127.6	53 I 126.9	54 Xe 131.3
55 Cs 132.9	56 Ba 137.3	57 La* 138.9	72 Hf 178.5	73 Ta 180.9	74 W 183.9	75 Re 186.2	76 Os 192.2	77 Ir 190.2	78 Pt 195.1	79 Au 197.0	80 Hg 200.5	81 Tl 204.4	82 Pb 207.2	83 Bi 209.0	84 Po (209)	85 At (210)	86 Rn (222)
87 Fr (223)	88 Ra (226)	89 Ac** (227)															

58 *Ce 140.1	59 Pr 140.9	60 Nd 144.2	61 Pm (145)	62 Sm 150.4	63 Eu 152.0	64 Gd 157.3	65 Tb 158.9	66 Dy 162.5	67 Ho 164.9	68 Er 167.3	69 Tm 168.9	70 Yb 173.0	71 Lu 175.0
90 **Th 232.0	91 Pa (231)	92 U (238)											

EACH BOX CONTAINS AN ELEMENT'S ATOMIC NUMBER, SYMBOL, AND ATOMIC WEIGHT. WEIGHTS ARE NOT WHOLE NUMBERS BECAUSE THEY REPRESENT AN AVERAGE OF SEVERAL ISOTOPES.

FOR A WONDERFULLY INFORMATION-RICH PERIODIC TABLE WITH A DETAILED PROFILE OF EVERY ELEMENT, SEE http://pearl1.lanl.gov/periodic/default.htm. ANOTHER WEB-BASED TABLE, AT www.colorado.edu/physics/2000/applets/a3.html, SHOWS THE ENERGIES OF ALL THE ELECTRONS IN EVERY ATOM.

WHAT'S SO PERIODIC ABOUT THE PERIODIC TABLE? WHAT PROPERTIES REPEAT THEMSELVES IN THE COLUMNS? WHAT TRENDS DO WE TRACE ALONG THE ROWS?

The Outermost Electrons

MOVING LEFT TO RIGHT ALONG A ROW OF MAIN-GROUP ELEMENTS, THE NUMBER OF OUTER ELEC-TRONS GOES UP STEADILY. GROUP 1 ELEMENTS ALL HAVE ONE OUTER ELECTRON, GROUP 2 ELEMENTS HAVE TWO, ETC., UNTIL THE LAST GROUP, WHICH ALL HAVE EIGHT. TRANSITION METALS HAVE EITHER ONE OR TWO OUTER ELECTRONS.*

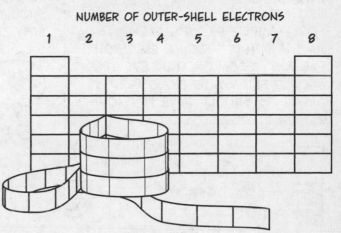

NUMBER OF OUTER-SHELL ELECTRONS

THE OUTER ELECTRONS, CALLED **VALENCE ELECTRONS**, ACCOUNT FOR MOST CHEMICAL REACTIONS.

Atomic Size

GOING ALONG A ROW FROM LEFT TO RIGHT, ATOMS GET SMALLER, AND MOVING DOWN A COLUMN, THEY GET BIGGER.

REASON: MOVING TO THE RIGHT, THE BIGGER CHARGE OF THE NUCLEUS PULLS ELECTRONS CLOSER IN. GOING DOWN A COLUMN, THE OUTER ELECTRONS ARE IN HIGHER SHELLS, HENCE FARTHER AWAY FROM THE NUCLEUS.

*TRANSITION METALS' INNER ELECTRONS SOMETIMES HAVE HIGH ENOUGH ENERGY TO ACT LIKE OUTER ELECTRONS, HOWEVER.

Ionization Energy

AN ATOM'S **IONIZATION ENERGY**—THE ENERGY NEEDED TO REMOVE AN OUTER ELECTRON—DEPENDS ON THE ATOM'S SIZE.

FOR EXAMPLE, GROUP 1 ELEMENTS HAVE A SINGLE VALENCE ELECTRON FAR AWAY FROM THE NUCLEUS. IT SHOULD BE EASY TO PRY OFF. THESE ELEMENTS SHOULD HAVE LOW IONIZATION ENERGIES.

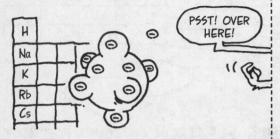

AND SO THEY DO. GROUP 1 ELEMENTS—LITHIUM, SODIUM, POTASSIUM, RUBIDIUM, AND CESIUM, THE **ALKALI METALS**—SHED ELECTRONS EASILY.

IN FACT, THEY ARE SO REACTIVE THAT THEY ARE NEVER FOUND NATURALLY PURE, BUT ALWAYS IN COMBINATION WITH OTHER ELEMENTS.

IN PURE FORM, THEY CAN BE EXPLOSIVELY DANGEROUS!

MOVING RIGHTWARD ALONG A ROW, ELECTRONS ARE CLOSER TO THE NUCLEUS, WHICH HOLDS THEM MORE TIGHTLY, SO IONIZATION ENERGIES SHOULD RISE TO A MAXIMUM IN THE LAST COLUMN.

AT THE START OF THE NEXT ROW, WITH A NEW OUTER SHELL, IONIZATION ENERGY DROPS AGAIN. THIS GRAPH SHOWS THE PERIODICITY OF IONIZATION ENERGY.

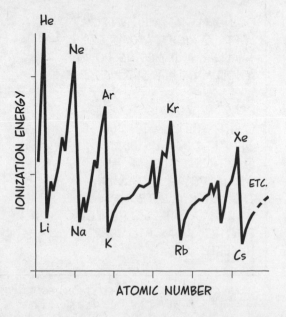

Electron Affinity

THIS PROPERTY, THE FLIP SIDE OF IONIZATION ENERGY, MEASURES AN ATOM'S "WILLINGNESS" TO BECOME AN ANION, I.E., TO ADD AN EXTRA ELECTRON.

STRAY ELECTRONS MAY FEEL THE NUCLEAR PULL AND ATTACH THEMSELVES TO ATOMS, ESPECIALLY IF AN UNFILLED OUTER ORBITAL IS AVAILABLE.

COME HEEERE, LITTLE ELECTRON!

HIGHER ELECTRON AFFINITY

						He
				O	F	Ne
				S	Cl	Ar
					Br	

HALOGENS

ATOMS TOWARD THE RIGHT SIDE OF THE PERIODIC TABLE TEND TO HAVE HIGHER ELECTRON AFFINITY: SMALL DIAMETER (SO ELECTRONS CAN GET CLOSER), BIG PULL FROM THE NUCLEUS, AND AN UNFILLED ORBITAL OR TWO.

EXCEPT IN THE LAST GROUP! THEY'RE FULL!

THE NEXT-TO-LAST GROUP IS ESPECIALLY ELECTRON HUNGRY. THESE ELEMENTS, THE **HALOGENS,** HAVE A SMALL DIAMETER AND ONE VACANT SPOT IN A p ORBITAL. AS YOU MIGHT IMAGINE, HALOGENS COMBINE WITH THE ELECTRON-SHEDDING ALKALI METALS OF GROUP 1. TABLE SALT, $NaCl$, IS A PRIME EXAMPLE OF AN ALKALI-HALOGEN COMPOUND.

THAT WAS **MY** ELECTRON, BUT I DON'T MIND...

THE PERIODIC TABLE IS BROADLY DIVIDED ALONG A STAIRSTEP BORDER INTO METALS AND NONMETALS, WITH A FEW CONFUSED "METALLOIDS" STRADDLING THE FENCE. METALS, ON THE LEFT, VASTLY OUTNUMBER NONMETALS, THANKS TO ALL THE ELEMENTS IN THE "LOOPS".

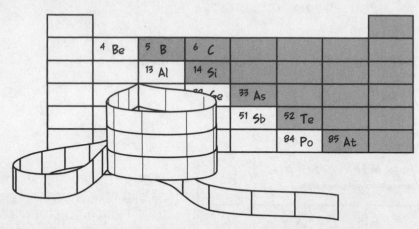

METALS TEND TO GIVE UP ELECTRONS FREELY, WHEREAS NONMETALS GENERALLY PREFER TO GAIN OR SHARE ELECTRONS. BUT METALS DO SHARE ELECTRONS AMONG THEMSELVES, FORMING TIGHTLY-PACKED, DENSE SOLIDS. NONMETALS USUALLY HAVE A LESS COHESIVE STRUCTURE.

Properties of metals

HIGH DENSITY

HIGH MELTING POINT AND BOILING POINT

GOOD ELECTRICAL CONDUCTIVITY

SHINY

MALLEABLE (EASY TO SHAPE)

DUCTILE (EASY TO STRETCH INTO WIRES)

REACTIVE WITH NONMETALS

Properties of nonmetals

OFTEN LIQUID OR GASEOUS AT ROOM TEMPERATURE

BRITTLE WHEN SOLID

DULL-LOOKING

POOR ELECTRICAL CONDUCTIVITY

REACTIVE WITH METALS (EXCEPT FOR THE LAST GROUP)

SOME NONMETALS BARELY REACT WITH **ANYTHING!**

DON'T YOU TOUCH ME!

| He |
| Ne |
| Ar |
| Kr |
| Xe |

THE LAST COLUMN OF THE PERIODIC TABLE IS UNIQUELY STRANGE. ITS DENIZENS, BECAUSE THEY LIVE FAR TO THE RIGHT, HAVE **HIGH IONIZATION ENERGIES,** SO THEY DON'T EASILY MAKE CATIONS. THEY ALSO HAVE **LOW ELECTRON AFFINITY** BECAUSE THEIR OUTER ORBITALS ARE FULL, SO THEY DON'T MAKE ANIONS EITHER!

THEY JUST... SIT THERE...

ALL EXCEPT HELIUM HAVE EIGHT OUTER ELECTRONS.

IN FACT, THEY RARELY REACT WITH ANYTHING. THEY JUST FLOAT AROUND IN AN UNCONNECTED, STANDOFFISH, GASEOUS STATE AND SO ARE KNOWN AS **NOBLE GASES.** YOU ALREADY KNOW ABOUT **NEON,** BUT THE MOST COMMON IS **ARGON** (ALMOST 1% OF THE ATMOSPHERE). IT IS USED IN ORDINARY INCANDESCENT LIGHT BULBS, SINCE IT WON'T REACT WITH THE HOT FILAMENT.

I NEED NOTHING. I YIELD NOTHING.

YES, PRINCESS ARGON!

JUST LIKE REAL NOBILITY, THE NOBLE GASES ARE THE ENVY OF THE COMMON ELEMENTS. EVERYONE WANTS THAT FULL COMPLEMENT OF EIGHT OUTER ELECTRONS.

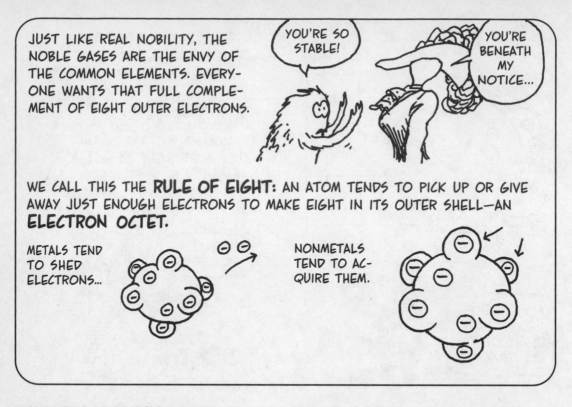

YOU'RE SO STABLE!

YOU'RE BENEATH MY NOTICE...

WE CALL THIS THE **RULE OF EIGHT:** AN ATOM TENDS TO PICK UP OR GIVE AWAY JUST ENOUGH ELECTRONS TO MAKE EIGHT IN ITS OUTER SHELL—AN **ELECTRON OCTET.**

METALS TEND TO SHED ELECTRONS...

NONMETALS TEND TO ACQUIRE THEM.

AND THIS BRINGS US TO THE SUBJECT OF OUR NEXT CHAPTER...

OO! IS THIS WHERE THEY GET EXPOSED TO WEIRD RAYS AND TURN INTO **RADIOACTIVE WEREWOLVES?**

UM... NOT EXACTLY...

BEFORE GOING ON, PLEASE TAKE A MOMENT TO APPRECIATE HOW AMAZING THIS CHAPTER HAS BEEN. STARTING FROM SOME WEIRD PROPERTIES OF ELEMENTARY ATOMIC PARTICLES, SCIENCE HAS MANAGED TO DESCRIBE THE ATOM, EXPLAIN THE PERIODIC TABLE, AND ACCOUNT FOR MANY CHEMICAL PROPERTIES OF THE ELEMENTS. NO WONDER ATOMIC THEORY HAS BEEN CALLED **"THE SINGLE MOST IMPORTANT IDEA IN SCIENCE."**

Chapter 3
Togetherness

IF ELEMENTS AND ATOMS WERE ALL THERE WERE, CHEMISTRY WOULD BE A PRETTY DULL SUBJECT. ATOMS WOULD JUST JIGGLE AROUND BY THEMSELVES LIKE A BUNCH OF NOBLE GASES, AND NOTHING WOULD HAPPEN.

LEAVE ME ALONE.

GLADLY.

BUT IN REALITY, CHEMISTRY IS A SORT OF FRENZY OF TOGETHERNESS. MOST ATOMS ARE GREGARIOUS LITTLE CRITTERS... AND THAT'S HOW WE'RE GOING TO DRAW THEM, SOMETIMES... AS LITTLE CRITTERS.

METAL

NONMETAL

THE COMBINATIONS ARE ENDLESS. METALS BOND TO METALS, NONMETALS TO NONMETALS, METALS TO NONMETALS. SOMETIMES ATOMS CLUMP TOGETHER IN LITTLE CLUSTERS AND SOMETIMES IN IMMENSE CRYSTAL ARRAYS. NO WONDER THE SUBJECT IS SO... SEXY!

I'M LEAVING YOU FOR POTASSIUM.

ATOMS COMBINE WITH EACH OTHER BY EXCHANGING OR SHARING ELECTRONS. THE DETAILS DEPEND ON THE PREFERENCES OF THE PARTICULAR ATOMS INVOLVED. DOES AN ATOM "WANT" TO SHED AN ELECTRON OR TO PICK ONE UP? AND HOW BADLY?

ELECTRONS! WHO NEEDS 'EM?

UM... ER... AH...

METALS, AS WE'VE SEEN, TEND TO GIVE UP ELECTRONS, THOUGH SOME METALS DO SO MORE EN- THUSIASTICALLY THAN OTHERS. A CHEMIST WOULD SAY THAT METALS ARE MORE OR LESS **ELECTROPOSITIVE.**

WHATEVER.

NONMETALS ARE MORE OR LESS **ELECTRONEGATIVE:** THEY TEND TO ACCEPT EXTRA ELECTRONS. SOME NONMETALS, LIKE FLUORINE AND OXYGEN, AVIDLY GRAB ELECTRONS, WHILE OTHERS, SUCH AS CARBON, CAN TAKE THEM OR LEAVE THEM.

IN BETWEEN ARE THE METALLOIDS, WHICH ARE COMPLETELY AMBIVALENT.

SIGH...

Ionic Bonds

WHEN A HIGHLY ELECTROPOSITIVE ATOM MEETS A HIGHLY ELECTRONEGATIVE ONE, THE RESULT IS AN IONIC BOND. THE ELECTROPOSITIVE ATOM EASILY GIVES AWAY ONE OR MORE ELECTRONS AND BECOMES A POSITIVELY CHARGED CATION. THE ELECTRONEGATIVE ATOM LOVES TO ACQUIRE EXTRA ELECTRONS AND IN DOING SO BECOMES AN ANION.

THE TWO IONS THEN EXPERIENCE AN ELECTROSTATIC ATTRACTION.

IN FACT, THEY ATTRACT NOT ONLY EACH OTHER, BUT EVERY OTHER CHARGED PARTICLE IN THE NEIGHBORHOOD.

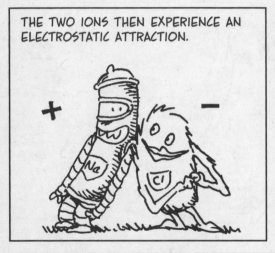

THEIR MUTUAL ATTRACTION PACKS THEM TOGETHER IN A DENSE, REGULAR **IONIC CRYSTAL.** IN THE CASE OF SODIUM AND CHLORIDE,* EACH ION HAS A SINGLE CHARGE SO NEUTRALITY IS ACHIEVED BY THIS SIMPLE CUBIC ARRANGEMENT:

HM... DID THIS GET OUT OF HAND OR WHAT?

YES... I CAN'T MOVE.

IF YOU LOOK CLOSELY AT TABLE SALT, YOU CAN SEE THAT THE CRYSTALS ARE LITTLE CUBES—EACH ONE A MONSTER ARRAY OF SODIUM AND CHLORIDE IONS.

*SINGLE-ATOM ANIONS ARE NAMED BY ADDING "IDE" TO THE ROOT OF THEIR ELEMENTAL NAME: FLUORIDE, OXIDE, ETC.

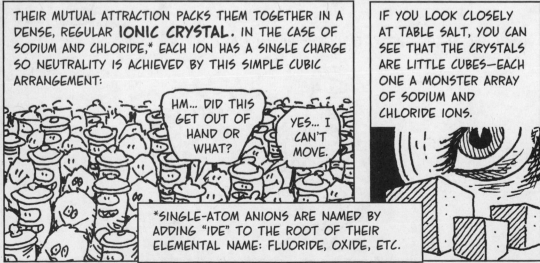

OTHER IONS MAY FORM DIFFERENT CRYSTALLINE STRUCTURES. WHEN CALCIUM, WHICH GIVES UP TWO ELECTRONS, COMBINES WITH CHLORINE, WHICH ACCEPTS ONLY ONE, TWO CHLORIDE IONS ARE NEEDED TO NEUTRALIZE EACH CALCIUM. WE WRITE AN ION WITH ITS ELEMENT SYMBOL AND CHARGE. SO THE CALCIUM ION IS Ca^{2+}, AND CHLORIDE IS Cl^-.

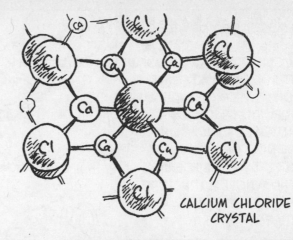

CALCIUM CHLORIDE CRYSTAL

THE FORMULA OF THESE IONIC CRYSTALS IS GIVEN "IN LOWEST TERMS." EVEN THOUGH A SODIUM CHLORIDE CRYSTAL MAY CONTAIN TRILLIONS OF ATOMS, WE WRITE ITS **EMPIRICAL FORMULA** AS NaCl. THIS SHOWS THAT THE CRYSTAL HAS ONE SODIUM ION FOR EACH CHLORIDE. IN THE SAME WAY, CALCIUM CHLORIDE IS WRITTEN $CaCl_2$.

OTHERWISE, THERE'S NO HOPE!

2 TRILLION, 6 HUNDRED BILLION, NINETY-**SIX**...

OCCASIONALLY, IONICALLY BONDED ATOMS HAVE NO NATURAL CRYSTALLINE ARRANGEMENT. INSTEAD THEY CLUMP TOGETHER INTO SMALL GROUPS CALLED **MOLECULES.** BORON TRIFLUORIDE, BF_3, IS AN IONIC COMPOUND THAT IS GASEOUS AT ROOM TEMPERATURE.

FACE FACTS—WE'RE A BAD FIT...

FINE. GOOD-BYE.

FORGET IT...

49

SOME IONS CONSIST OF MORE THAN ONE ATOM. WE'LL SEE HOW TO BUILD THESE **POLYATOMIC** IONS LATER IN THE CHAPTER. THESE THINGS BEHAVE VERY MUCH LIKE MONOATOMIC IONS, EXCEPT FOR THEIR SHAPE. THE WHOLE STRUCTURE ACTS AS A SINGLE CHARGED UNIT.

COME TO DA-DA!

A TYPICAL EXAMPLE IS **SULFATE**, SO_4^{2-}, AN ANION THAT BONDS WITH Ca^{2+} TO MAKE **CALCIUM SULFATE**, $CaSO_4$, AN INGREDIENT OF WALLBOARD.

OTHER EXAMPLES:

NH_4^+	AMMONIUM
OH^-	HYDROXIDE
NO_2^-	NITRITE
NO_3^-	NITRATE
HCO_3^-	BICARBONATE
CO_3^{2-}	CARBONATE
SO_3^{2-}	SULFITE
PO_4^{3-}	PHOSPHATE

EACH POLYATOMIC ION MUST BE REGARDED AS A SINGLE ION. FOR EXAMPLE, ALUMINUM HYDROXIDE, WHICH COMBINES Al^{3+} AND OH^-, MUST HAVE THREE HYDROXIDES TO BALANCE EACH ALUMINUM. THE FORMULA IS WRITTEN $Al(OH)_3$, AND THE CRYSTAL STRUCTURE LOOKS LIKE THIS:

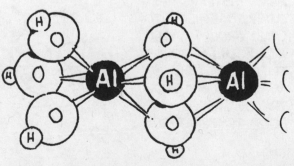

IONIC BONDS ARE STRONG. IT TAKES A LOT OF ENERGY TO BREAK THEM. THIS EXPLAINS WHY MOST IONIC CRYSTALS HAVE SUCH HIGH MELTING POINTS: TREMENDOUS HEAT IS NEEDED TO JAR THE IONS LOOSE AND GET THEM SLOSHING AROUND AS A LIQUID.

TABLE SALT MELTS AT 801° C.

AND YET—HIT A SALT CRYSTAL WITH A HAMMER AND IT CRUMBLES. WHY SHOULD IT BE SO BRITTLE?

WHY SHOULD I HIT SALT WITH A HAMMER?

ANSWER: WHEN WHACKED, THE CRYSTAL MAY DEVELOP TINY CRACKS, AND ONE LAYER MAY SHIFT SLIGHTLY ACROSS ANOTHER.

THIS SHIFT CAN ALIGN POSITIVES OPPOSITE POSITIVES AND NEGATIVES OPPOSITE NEGATIVES. NOW THE TWO CHUNKS REPEL EACH OTHER, AND THE CRYSTAL LITERALLY FLIES APART.

CRACK

BUT NOT ALL CRYSTALS BEHAVE THIS WAY—METALLIC CRYSTALS, FOR EXAMPLE...

Metallic Bonds

PURE METALS ALSO FORM CRYSTALS, THOUGH YOU PROBABLY DON'T THINK OF THEM THAT WAY. THEY LACK THE TRANSPARENCY AND SPARKLE OF NaCl AND OTHER IONIC CRYSTALS, AND THEY USUALLY AREN'T BRITTLE.

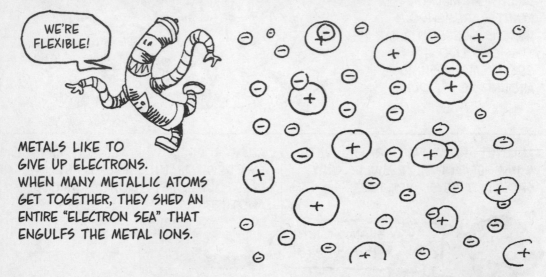

WE'RE FLEXIBLE!

METALS LIKE TO GIVE UP ELECTRONS. WHEN MANY METALLIC ATOMS GET TOGETHER, THEY SHED AN ENTIRE "ELECTRON SEA" THAT ENGULFS THE METAL IONS.

PULLED FROM ALL DIRECTIONS, THE METAL IONS FIND IT HARD TO MOVE, AND THEY PACK TIGHTLY TOGETHER IN CRYSTALLINE STRUCTURES. THERE ARE SEVERAL POSSIBLE PACKING ARRANGEMENTS, ALL OF THEM DENSE. HERE ARE TWO.

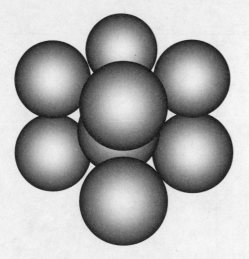

BODY-CENTERED CUBIC: EACH ATOM SURROUNDED BY EIGHT OTHERS

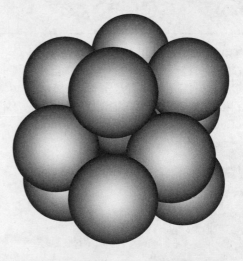

FACE-CENTERED CUBIC: EACH ATOM SURROUNDED BY TWELVE OTHERS

METALS TEND TO BE GOOD CONDUCTORS OF ELECTRICITY. THE LIGHT, FREE ELECTRONS MOVE AROUND EASILY. NEGATIVE CHARGE COMING FROM OUTSIDE CAN PUSH THE "SEA" OF ELECTRONS, MAKING A CURRENT.

LIKE ANY CRYSTAL, BEING WHACKED BY A HAMMER MAY CAUSE A METAL'S CRYSTALLINE STRUCTURE TO CRACK AND SHIFT.

BUT UNLIKE IONIC CRYSTALS, THE METAL'S IONIC REPULSION IS OVERCOME BY THAT NEGATIVE SEA OF ELECTRONS HOLDING ALL THE ATOMS IN PLACE.

WHO'D HAVE THOUGHT ELECTRONS WOULD HAVE A CALMING EFFECT?

SO, INSTEAD OF SHATTERING, A METAL TENDS TO BEND OR STRETCH.*

IT'S LIKE MAGIC!

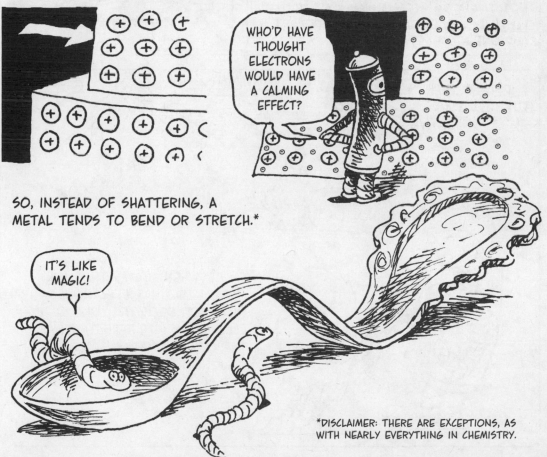

*DISCLAIMER: THERE ARE EXCEPTIONS, AS WITH NEARLY EVERYTHING IN CHEMISTRY.

Covalent Bonding and Molecules

I JUST CAN'T BE BOTHERED.

GIMME GIMME!

METALLIC BONDING HAPPENS WHEN A LOT OF ELECTROPOSITIVE ATOMS ARE TRAPPED BY ALL THE ELECTRONS THEY SHARE. IT'S LIKE A COMMUNAL HOUSEHOLD.

YOU CAN'T ESCAPE ELECTRONS...

IONIC BONDS FORM WHEN A HIGHLY ELECTRONEGATIVE ATOM MEETS A HIGHLY ELECTROPOSITIVE ONE. ELECTRONS ARE HANDED OFF, AND ONE ATOM GETS SOLE CUSTODY.

AND THEN THERE'S EVERYTHING ELSE: THE BONDS BETWEEN TWO ELECTRONEGATIVE ATOMS...

GIMME

GIMME

HERE... NO... UM...

OR BETWEEN ATOMS THAT ARE ONLY SOMEWHAT ELECTRONEGATIVE OR ELECTROPOSITIVE. ONE SHEDS ELECTRONS, BUT RELUCTANTLY... THE OTHER ACCEPTS THEM, BUT HALF-HEARTEDLY... AND THE RESULT IS A SORT OF MARRIAGE, OR JOINT CUSTODY ARRANGEMENT.

UNPAIRED ELECTRON— BAD!

THE SIMPLEST POSSIBLE EXAMPLE IS HYDROGEN. A LONE HYDROGEN ATOM HAS AN UNPAIRED ELECTRON, WHICH THE ATOM CAN EITHER GIVE UP OR PAIR WITH ANOTHER ELECTRON.

WHEN ONE HYDROGEN ENCOUNTERS ANOTHER, THEIR ELECTRONS NATURALLY PAIR UP IN A SINGLE, SHARED ORBITAL.

THIS PAIR PULLS ON BOTH NUCLEI, SO IT HOLDS THE ATOMS TOGETHER. THE BOND IS CALLED **COVALENT**, BECAUSE BOTH ATOMS CONTRIBUTE EQUALLY.

EACH HYDROGEN ATOM "THINKS" IT HAS A FULL 1s VALENCE SHELL, SO THE RESULTING TWOSOME, OR HYDROGEN **MOLECULE**, H_2, IS STABLE.

AT NORMAL TEMPERATURES, HYDROGEN GAS IS ALWAYS IN MOLECULAR FORM!

MORE EXAMPLES: OXYGEN, THE SECOND-MOST ELECTRONEGATIVE ELEMENT (AFTER FLUORINE), HAS SIX VALENCE ELECTRONS. WE INDICATE THIS WITH A "LEWIS DIAGRAM" THAT REPRESENTS EACH OF THESE OUTER ELECTRONS AS A DOT.

AGH! I SMELL RUST!

WHEN TWO OXYGENS GET TOGETHER, THEY BOND COVALENTLY BY SHARING FOUR ELECTRONS, AS SHOWN IN THIS LEWIS DIAGRAM:

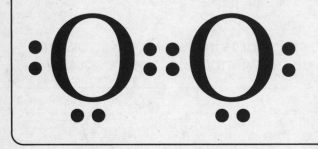

HERE, TOO, BOTH ATOMS NOW HAVE A FULL OUTER OCTET. (COUNT THE ELECTRONS!) WHEN FOUR ELECTRONS ARE SHARED IN THIS WAY, WE CALL IT A **DOUBLE BOND** AND SOMETIMES WRITE IT AS O=O.

NITROGEN, WITH FIVE VALENCE ELECTRONS, FORMS TRIPLE COVALENT BONDS TO MAKE N_2 OR N≡N.

THE ATMOSPHERE IS MOSTLY N_2 AND O_2.

MANY OTHER NON-METALS, INCLUDING THE HALOGENS, FORM DIATOMIC (TWO-ATOM) MOLECULES IN THIS WAY.

COVALENT BONDING INVOLVES ELECTRON SHARING BETWEEN A SPECIFIC PAIR OF ATOMS. IT'S LIKE A HANDSHAKE.

ONLY YOOOU...

SINCE ATOMS HAVE ONLY A LIMITED NUMBER OF "HANDS," COVALENT COMPOUNDS ARE USUALLY FOUND IN THE FORM OF **MOLECULES**, OR SMALL, DISCRETE GROUPS OF ATOMS.

CO_2

EVERY MOLECULE IN A PURE SUBSTANCE HAS THE SAME COMPOSITION. WE WRITE ITS FORMULA ACCORDING TO THE NUMBER OF EACH KIND OF ATOM PRESENT.

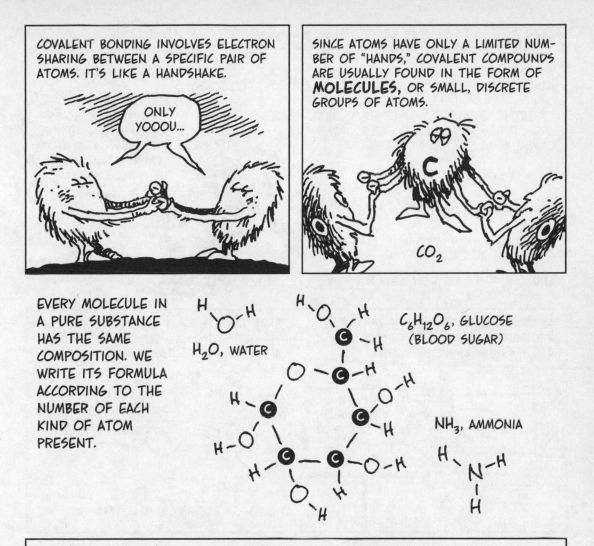

H_2O, WATER

$C_6H_{12}O_6$, GLUCOSE (BLOOD SUGAR)

NH_3, AMMONIA

OCCASIONALLY WE DO SEE COVALENTLY BONDED CRYSTALS. DIAMOND, FOR EXAMPLE, CONSISTS OF A SO-CALLED **COVALENT NETWORK** OF CARBON ATOMS.

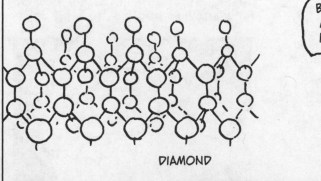

DIAMOND

BY THE WAY, IS AN **ONION** A KIND OF ION?

YOUR PUN MAKES ME WEEP...

Molecular Shapes

SO FAR, WE'VE LOOKED ONLY AT COVALENT BONDS BETWEEN TWO IDENTICAL ATOMS. NOW LET'S SEE HOW DIFFERENT ATOMS CAN SHARE ELECTRONS.

CARBON DIOXIDE, FAMOUS EXHAUST GAS, CO_2: CARBON HAS FOUR VALENCE ELECTRONS AND OXYGEN HAS SIX, SO WE WRITE:

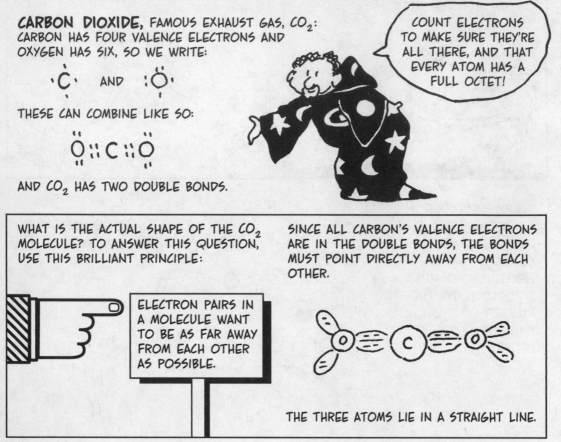

$\cdot\overset{\cdot}{\underset{\cdot}{C}}\cdot$ AND $:\overset{\cdot\cdot}{\underset{\cdot\cdot}{O}}\cdot$

THESE CAN COMBINE LIKE SO:

$\overset{\cdot\cdot}{\underset{\cdot\cdot}{O}}::C::\overset{\cdot\cdot}{\underset{\cdot\cdot}{O}}$

AND CO_2 HAS TWO DOUBLE BONDS.

COUNT ELECTRONS TO MAKE SURE THEY'RE ALL THERE, AND THAT EVERY ATOM HAS A FULL OCTET!

WHAT IS THE ACTUAL SHAPE OF THE CO_2 MOLECULE? TO ANSWER THIS QUESTION, USE THIS BRILLIANT PRINCIPLE:

ELECTRON PAIRS IN A MOLECULE WANT TO BE AS FAR AWAY FROM EACH OTHER AS POSSIBLE.

SINCE ALL CARBON'S VALENCE ELECTRONS ARE IN THE DOUBLE BONDS, THE BONDS MUST POINT DIRECTLY AWAY FROM EACH OTHER.

THE THREE ATOMS LIE IN A STRAIGHT LINE.

IN **SULFUR TRIOXIDE,** SO_3, SULFUR AND OXYGEN EACH HAVE SIX VALENCE ELECTRONS.

$:\overset{\cdot\cdot}{\underset{\cdot}{S}}\cdot$ $\cdot\overset{\cdot\cdot}{\underset{\cdot}{O}}:$

THREE OXYGENS CAN BOND TO SULFUR.

$:\overset{\cdot\cdot}{\underset{}{O}}:$

$:\overset{\cdot\cdot}{\underset{\cdot\cdot}{O}}::\overset{}{S}:\overset{\cdot\cdot}{\underset{\cdot\cdot}{O}}:$

(THE DOUBLE BOND COULD GO ON ANY ONE OF THE OXYGENS.)

USING THE PRINCIPLE THAT ELECTRON PAIRS MUST AVOID EACH OTHER (EXCEPT FOR THE ONES IN THE DOUBLE BOND— THEY'RE STUCK), WE CONCLUDE THAT SO_3 IS TRIANGULAR AND LIES IN A PLANE.

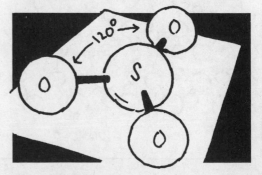

CARBON TETRACHLORIDE, CCl_4, AN INDUSTRIAL SOLVENT, COMBINES

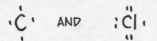

$\cdot \ddot{C} \cdot$ AND $: \ddot{C}l \cdot$

WITH FOUR SINGLE BONDS.

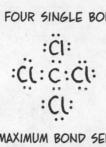

FOR MAXIMUM BOND SEPARATION, THIS MOLECULE HAS A TETRAHEDRAL SHAPE, WITH THE OUTER ATOMS AT THE POINTS OF A TRIANGULAR PYRAMID.

AMMONIA, NH_3. YOU MIGHT EXPECT THIS TO BE A TRIANGLE, BUT THE LEWIS DIAGRAM SAYS OTHERWISE. THE FOURTH ELECTRON PAIR REPELS THE OTHERS, AND WE GET A TETRAHEDRON WITH H AT THREE OF THE VERTICES.

EXTRA PAIR

WATER, H_2O, IS SIMILAR. IT HAS TWO ELECTRON PAIRS WITH NOTHING ATTACHED TO THEM. THEY, TOO, MUST BE TAKEN INTO ACCOUNT.

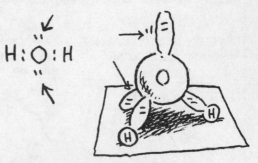

MOLECULES LIKE NH_3 AND H_2O ARE CALLED **BENT.**

THIS COVERS THE SHAPES OF THE MOST COMMON MOLECULES, ALTHOUGH THERE ARE SOME ODDITIES LIKE SF_6, WHERE THE SULFUR HAS SIX ELECTRON PAIRS.

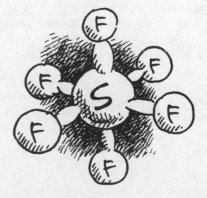

SF_6 IS OCTAHEDRAL.

Shape and Orbital Bond Theory (advanced)

ON THE PREVIOUS TWO PAGES, WE USED THE PRINCIPLE THAT ELECTRON PAIRS IN MOLECULES STAY AWAY FROM EACH OTHER. WE CAN ACCOUNT FOR THIS FACT IN TERMS OF ELECTRON ORBITALS.

WHEN H BONDS WITH H, TWO s ORBITALS MERGE. THIS IS CALLED A σ (SIGMA) BOND.

IN O_2, TWO ELECTRONS IN p ORBITALS ARE SHARED IN A π (PI) BOND.

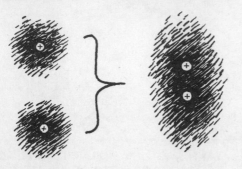

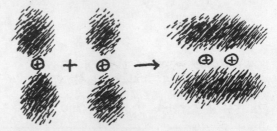

(WE HAVE OMITTED THE NONBONDING ORBITALS.)

BUT IN GENERAL, WE GET SOMETHING CALLED **HYBRID ORBITALS.** FOR EXAMPLE:

CARBON, WITH $2s^2 2p^2$, HAS TWO PAIRED s ELECTRONS AND TWO UNPAIRED p ELECTRONS.

WHEN A HYDROGEN ATOM APPROACHES, ITS NUCLEUS PULLS ON C'S ELECTRONS, RAISING THEIR ENERGY.

ONE s ELECTRON IS "PROMOTED" TO A p ORBITAL, AND NOW ALL ARE UNPAIRED.

THE UNPAIRED ORBITALS "HYBRIDIZE" AND BECOME LOPSIDED. SUCH AN ORBITAL IS CALLED AN sp **HYBRID.** ONE OF THEM LOOKS LIKE THIS.

AND FOUR OF THEM LOOK LIKE THIS. (HERE EACH ONE IS BONDED TO A HYDROGEN ATOM.)

THE LOPSIDED LOBES MUST REPEL EACH OTHER, SO THE CH_4 MOLECULE MUST BE A TETRAHEDRON. **THE MOLECULE'S GEOMETRY IS CAUSED BY THE SHAPE OF HYBRID ORBITALS.**

More on Lewis Diagrams and Charged Molecules

IN A LEWIS DIAGRAM, EACH ATOM ENDS UP WITH A COMPLETE OCTET (USUALLY—SEE BELOW). THIS CAN OFTEN HAPPEN IN MORE THAN ONE WAY. FOR INSTANCE, WE JUST SAW SO_3, BUT SO_2 ALSO EXISTS, AND IS ACTUALLY THE MORE COMMON OXIDE OF SULFUR.

NOTE UNBONDED PAIR

SULFUR'S EXTRA ELECTRON PAIR IMPLIES THAT THE MOLECULE IS BENT.

INCIDENTALLY, THE DOUBLE BOND ISN'T REALLY ON ONE OXYGEN OR THE OTHER, BUT SOMEHOW HALFWAY ON BOTH AT THE SAME TIME, A QUANTUM-MECHANICAL MYSTERY KNOWN AS **RESONANCE.**

$$O=S-O \rightleftharpoons O-S=O$$

WE CAN ALSO WRITE A LEWIS DIAGRAM FOR **SULFATE**, SO_4^{2-}, WITH NO DOUBLE BONDS AT ALL. THIS LOOKS NICE AND NATURAL, EXCEPT THAT TWO EXTRA ELECTRONS ARE REQUIRED TO COMPLETE ALL THE BONDS. SO SO_4^{2-} IS REALLY A COVALENTLY BONDED POLYATOMIC ION WITH A CHARGE OF −2.

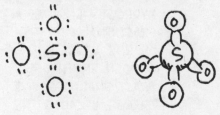

MORE POLYATOMIC IONS:

NITRATE, NO_3^-, HAS ONE EXTRA ELECTRON AND RESONANCE BETWEEN THREE DIFFERENT FORMS.

$$O=N-O \rightleftharpoons O-N-O \rightleftharpoons O-N=O$$

HYDROXIDE, OH^-, HAS ONE EXTRA ELECTRON.

USUALLY, ALL ELECTRONS ARE PAIRED AND EVERY ATOM GETS A FULL OCTET—BUT THERE ARE EXCEPTIONS. IN NITROGEN DIOXIDE, NO_2, NITROGEN HAS AN UNPAIRED ELECTRON.

AND IN BERYLLIUM FLUORIDE, BeF_2, Be GETS ONLY HALF AN OCTET.

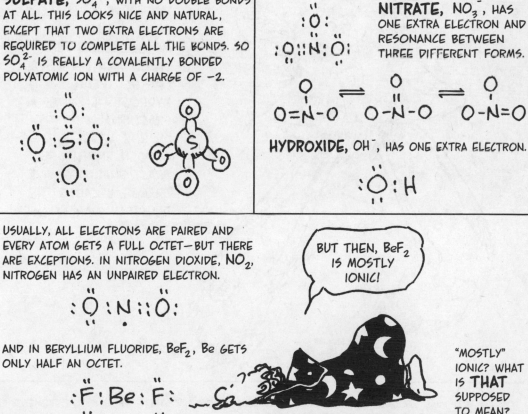

BUT THEN, BeF_2 IS MOSTLY IONIC!

"MOSTLY" IONIC? WHAT IS **THAT** SUPPOSED TO MEAN?

Polarity

MANY BONDS ARE NOT PURELY
COVALENT OR IONIC, BUT
SOMEWHERE IN BETWEEN.

CONSIDER WATER, H_2O. OXYGEN, WITH AN ELECTRONEGATIVITY VALUE (EN) OF 3.5, IS MORE ELECTRONEGATIVE THAN HYDROGEN (EN = 2.1).* THIS MEANS THAT THE ELECTRONS IN THE O—H BOND ARE NOT EQUALLY SHARED, BUT TEND TO HOVER CLOSER TO THE OXYGEN ATOM.

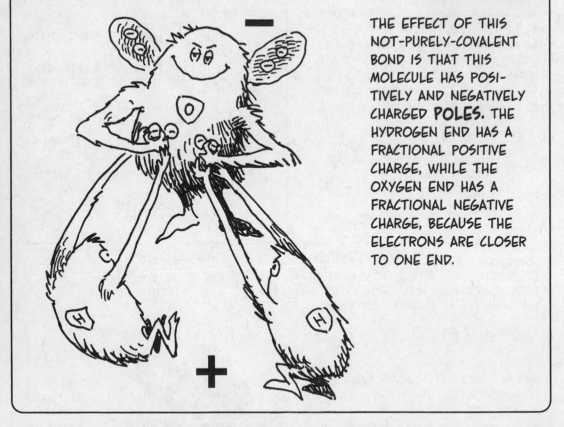

THE EFFECT OF THIS NOT-PURELY-COVALENT BOND IS THAT THIS MOLECULE HAS POSI-TIVELY AND NEGATIVELY CHARGED **POLES.** THE HYDROGEN END HAS A FRACTIONAL POSITIVE CHARGE, WHILE THE OXYGEN END HAS A FRACTIONAL NEGATIVE CHARGE, BECAUSE THE ELECTRONS ARE CLOSER TO ONE END.

*ON AN ARTIFICIAL SCALE RANGING FROM **0.7** FOR CESIUM, THE MOST ELECTROPOSITIVE ELEMENT, TO **4.0** FOR FLUORINE, THE MOST ELECTRONEGATIVE.

A BOND LIKE O—H, IN WHICH THE ELECTRONS ARE CLOSER TO ONE END, IS CALLED **POLAR.** POLAR BONDS ARE INTERMEDIATE BETWEEN COVALENT BONDS (EQUAL SHARING) AND IONIC BONDS (COMPLETE TRANSFER OF ELECTRONS).

| IONIC | STRONGLY POLAR | WEAKLY POLAR | COVALENT |

THE POLARITY OF BONDS AFFECTS THE WAY CHARGE IS DISTRIBUTED OVER A MOLECULE.

A BOND'S POLARITY DEPENDS ON THE DIFFERENCE IN ELECTRONEGATIVITY BETWEEN TWO ATOMS. BIGGER DIFFERENCES MEAN MORE POLARITY, WITH A DIFFERENCE OF **2.0** OR MORE BEING CONSIDERED IONIC.

BOND	EN DIFF.	BOND TYPE
N≡N	0	COVALENT
C—H	0.4	ESSENTIALLY COVALENT
O—H	1.4	MODERATELY POLAR
H—F	1.9	STRONGLY POLAR
Li—F	3.0	IONIC

SAMPLE ELECTRONEGATIVITIES			
H	2.1	Na	0.9
Li	1.0	Mg	1.2
C	2.5	S	2.5
N	3.0	Cl	3.0
O	3.5	K	0.8
F	4.0	Ca	1.0

THE POLARITY OF WATER EXPLAINS SOME OF ITS FAMILIAR PROPERTIES. FOR INSTANCE:

WATER IS LIQUID AT ROOM TEMPERATURE. THE PARTIAL CHARGES AT EACH END OF A WATER MOLECULE MAKE THE MOLECULES ATTRACT EACH OTHER, END TO END. WATER BONDS WEAKLY TO ITSELF. THIS INTERNAL COHESION HOLDS WATER TOGETHER IN LIQUID FORM.

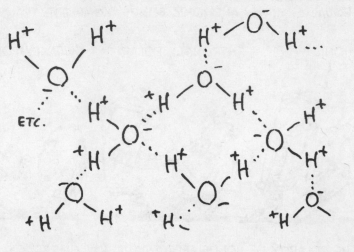

BY CONTRAST, THE MUCH HEAVIER BUT LESS POLAR SO_2 HAS LITTLE MUTUAL ATTRACTION, SO IT FORMS A GAS AT ROOM TEMPERATURE.

POLARITY ALSO EXPLAINS WHY WATER IS SO GOOD AT DISSOLVING IONIC COMPOUNDS SUCH AS TABLE SALT. THE CRYSTAL'S IONIC BONDS SLOWLY GIVE WAY TO THE PULL OF WATER'S POLES, AS IONS BREAK OFF THE CRYSTAL AND ATTACH THEMSELVES TO WATER MOLECULES.

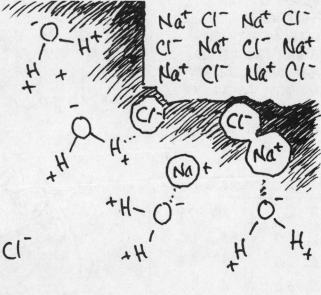

SIMILARLY, THE WEAK ATTRACTION OF A POLAR H TO ANOTHER MOLECULE IS CALLED **HYDROGEN BONDING.** IT HAPPENS TO BE A KEY FEATURE OF THE CHEMISTRY OF LIFE (SEE PAGE 241).

IONIC, COVALENT, METALLIC: THESE ARE THE MAIN TYPES OF CHEMICAL BONDS. WE'VE SEEN HOW THESE INTERATOMIC INTERACTIONS ARISE FROM THE ELECTRICAL PROPERTIES OF ATOMS, AND HOW THEY AFFECT THE STRUCTURES OF SUBSTANCES. NOW WE WANT TO FIND OUT WHAT THEY HAVE TO DO WITH THE CHEM...

HM?

EXCUSE ME!!

LOOK—ALL THIS INVISIBLE SUB-ATOMIC STUFF IS INTERESTING AND ALL... BUT ISN'T THIS BOOK SUPPOSED TO BE ABOUT CHEMISTRY?

UH... YES...?

SO WHERE ARE THE EXPLOSIONS? THE BUBBLING, NOXIOUS GOO? WHEN CAN I SMELL SOMETHING VILE? COME ON!!!

ER... COUGH... THE SECRETS OF THE UNIVERSE AREN'T ENOUGH FOR YOU?

WELL, I GUESS NOT!

AH... WELL THEN... I THINK YOU MAY LIKE THE NEXT CHAPTER...

Chapter 4
Chemical Reactions

OOPS! SOMEHOW WE FIND OURSELVES MAROONED ON A DESERT ISLAND. HOW
ARE WE GOING TO SURVIVE? MAYBE WE CAN MAKE SOMETHING USEFUL OUT
OF THE MATERIALS AT HAND...

Combustion, Combination, Decomposition

ALL WE NEED IS DRY WOOD AND OXYGEN!

LET'S WRITE A **REACTION EQUATION** FOR FIRE. WOOD CONTAINS MANY DIFFERENT MATERIALS, BUT IT'S MAINLY MADE OF C, H, AND O IN THE RATIO 1:2:1. WE CAN WRITE THE EMPIRICAL FORMULA FOR WOOD AS CH_2O, AND THEN FIRE LOOKS LIKE THIS:*

$$CH_2O \ (s) + O_2 \ (g) \longrightarrow CO_2 \ (g)\uparrow + H_2O \ (g)\uparrow$$

THE NOTATION EXPLAINED: THE SUBSTANCES ON THE LEFT OF THE HORIZONTAL ARROW $\longrightarrow$ ARE CALLED **REACTANTS**. ON THE RIGHT ARE THE **REACTION PRODUCTS**. $\xrightarrow{\Delta}$ WILL MEAN THAT HEAT WAS ADDED. THE SMALL LETTERS IN PARENTHESES SHOW THE PHYSICAL STATE OF THE CHEMICALS: g = GAS; s = SOLID; l = LIQUID; aq = DISSOLVED IN WATER. $\uparrow$ MEANS AN ESCAPING GAS, AND $\downarrow$ WILL MEAN A SOLID SETTLING OUT OF SOLUTION, OR **PRECIPITATING**.

SO OUR EQUATION READS: SOLID WOOD PLUS GASEOUS OXYGEN AND HEAT MAKES GASEOUS CARBON DIOXIDE PLUS WATER VAPOR. THIS IS A TYPICAL **COMBUSTION REACTION**. (YOU CAN TEST FOR THE WATER BY HOLDING A COOL GLASS OVER THE FLAME; DROPLETS WILL CONDENSE ON IT.)

*WE'RE LEAVING OUT PARTIALLY OR WHOLLY NONCOMBUSTED PRODUCTS SUCH AS SOOT, SMOKE, CO, ETC.

NOW THAT WE HAVE FIRE, WE'LL MAKE A BETTER FUEL: **CHARCOAL.** WE PUT DRY WOOD AND COCONUT SHELLS IN A PIT (TO LIMIT AVAILABLE OXYGEN) AND FIRE IT UP. THE REACTION IS*

$$CH_2O \xrightarrow{\Delta} C(s) + H_2O(g)\uparrow$$

THIS IS A **DECOMPOSITION** REACTION (OF THE FORM AB → A + B). IT MAKES ELEMENTAL CARBON, OR CHARCOAL.

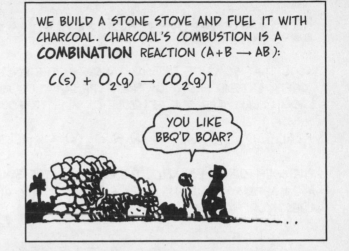

WE BUILD A STONE STOVE AND FUEL IT WITH CHARCOAL. CHARCOAL'S COMBUSTION IS A **COMBINATION** REACTION (A+B → AB):

$$C(s) + O_2(g) \rightarrow CO_2(g)\uparrow$$

YOU LIKE BBQ'D BOAR?

IN THIS OVEN WE CAN MAKE **POTTERY.** WE SCOOP A FINE-GRAINED MINERAL, KAOLINITE, FROM THE LAKE BOTTOM AND GRIND IT WITH A LITTLE WATER TO MAKE A SMOOTH KAOLIN CLAY, $Al_2Si_2O_5(OH)_4$. WE SHAPE THIS INTO VESSELS AND FIRE THEM IN A HOT OVEN:

$$3Al_2Si_2O_5(OH)_4(s) \xrightarrow{\Delta} Al_6Si_2O_{13}(s) + 4SiO_2(s) + 6H_2O(g)\uparrow$$

THE FIRST PRODUCT IS CALLED **MULLITE.** THE SECOND, SiO_2, IS **SILICA,** OR SAND—AND MELTED, IT'S **GLASS.** WHEN THE CLAY IS FIRED, MULLITE FUSES WITH THE GLASSY SILICA TO FORM A VERY HARD, WATERPROOF POT.

*MORE OR LESS. AGAIN WE IGNORE TRACE REACTANTS AND PRODUCTS.

Balancing Equations

NOTE THAT SOME OF THE SUBSTANCES IN THE POTTERY REACTION HAVE NUMERICAL COEFFICIENTS IN FRONT OF THEM. THE EQUATION MEANS THREE MOLECULES OF KAOLIN CLAY YIELD ONE MOLECULE OF MULLITE, FOUR OF SILICA, AND SIX OF WATER.

$$3Al_2Si_2O_5(OH)_4 \, (s) \xrightarrow{\triangle} Al_6Si_2O_{13} \, (s) + 4SiO_2 \, (s) + 6H_2O \, (g)\uparrow$$

THE COEFFICIENTS **BALANCE** THE EQUATION. THE SAME NUMBER OF EACH KIND OF ATOM APPEARS ON BOTH SIDES: 6 Al, 6 Si, 27 O, AND 12 H. HOW DO WE FIND THESE COEFFICIENTS?

START WITH AN UNBALANCED EQUATION

$$Al_2Si_2O_5(OH)_4 \, (s) \xrightarrow{\triangle} Al_6Si_2O_{13} \, (s) + SiO_2 \, (s) + H_2O \, (g)\uparrow$$

WRITE DOWN THE NUMBER OF ATOMS ON EACH SIDE.

	L	R
Al	2	6
Si	2	3
O	9	16
H	4	2

BALANCE ONE ELEMENT. WE START WITH Al.
MULTIPLY BY 3 ON THE LEFT TO GET:

$$3 \, Al_2Si_2O_5(OH)_4 \, (s) \xrightarrow{\triangle} Al_6Si_2O_{13} \, (s) + SiO_2 \, (s) + H_2O \, (g)\uparrow$$

AGAIN COUNT ATOMS ON EACH SIDE.

	L	R
Al	6	6
Si	6	3
O	27	16
H	12	2

BALANCE ANOTHER ELEMENT. WE CAN BALANCE Si BY PUTTING A 4 IN FRONT OF SiO_2:

$$3 \, Al_2Si_2O_5(OH)_4 \, (s) \xrightarrow{\triangle} Al_6Si_2O_{13} \, (s) + 4SiO_2 \, (s) + H_2O \, (g)\uparrow$$

AGAIN COUNT ATOMS ON EACH SIDE.

	L	R
Al	6	6
Si	6	6
O	27	22
H	12	2

FINALLY, A 6 IN FRONT OF H_2O BALANCES BOTH H AND O.

$$3 \, Al_2Si_2O_5(OH)_4 \, (s) \xrightarrow{\triangle} Al_6Si_2O_{13} \, (s) + 4SiO_2 \, (s) + 6H_2O \, (g)\uparrow$$

	L	R
Al	6	6
Si	6	6
O	27	27
H	12	12

GENERAL PROCEDURE:

- WRITE THE EQUATION WITHOUT COEFFICIENTS.
- LIST THE ELEMENTS IN THE EQUATION.
- CHECK THE NUMBER OF EACH KIND OF ATOM ON BOTH SIDES.
- BALANCE ATOMS ONE ELEMENT AT A TIME BY ADJUSTING COEFFICIENTS.
- REDUCE TO LOWEST TERMS IF NECESSARY.

THE ACT, OR ART, OF BALANCING EQUATIONS IS CALLED **REACTION STOICHIOMETRY.**

HERE ARE SOME PRACTICE EXAMPLES. SUPPLY COEFFICIENTS IN EACH EQUATION.

$$Al(s) + Fe_2O_3(s) \xrightarrow{\Delta} Al_2O_3(s) + Fe(s)$$

$$KClO_3(s) \xrightarrow{\Delta} KCl(s) + O_2(g)$$

$$C_4H_{10}(g) + O_2(g) \longrightarrow CO_2(g) + H_2O(g)$$

$$N_2(g) + H_2(g) \longrightarrow NH_3(g)$$

$$P_4(s) + F_2(g) \longrightarrow PF_5(g)$$

$$Zn(NO_3)_2(s) \xrightarrow{\Delta} ZnO(s) + NO_2(g) + O_2(g)$$

$$H_3PO_4(l) \xrightarrow{\Delta} H_2O(l) + P_4O_{10}(s)$$

$$Cu(s) + AgNO_3(aq) \longrightarrow Cu(NO_3)_2(aq) + Ag\downarrow$$

$$Fe(s) + O_2(g) \longrightarrow Fe_2O_3(s)$$

$$FeCl_3(s) + H_2O(l) \longrightarrow HCl(aq) + Fe(OH)_3\downarrow$$

The Mole

BALANCE IS A BEAUTIFUL THING!

?

THE EQUATION'S COEFFICIENTS LET US FIND THE RELATIVE **MASSES** OF PRODUCTS AND REACTANTS. THE CALCULATION USES A UNIT CALLED THE **MOLE.** ONE MOLE OF A SUBSTANCE IS THE AMOUNT WHOSE MASS EQUALS THE MOLECULAR OR ATOMIC WEIGHT OF THE SUBSTANCE **EXPRESSED IN GRAMS.**

THAT'S KIND OF A MOUTHFUL FOR A SIMPLE IDEA. LET'S ILLUSTRATE BY EXAMPLE:

	"MOLECULAR" WEIGHT	MOLAR WEIGHT
O_2	32 AMU	32 GRAMS
SiO_2	60 AMU	60 GRAMS
$Al_2Si_2O_5(OH)_4$	258 AMU	258 GRAMS
Fe	56 AMU	56 GRAMS
PROTON	1 AMU	1 GRAM
NaCl	58.5 AMU	58.5 GRAMS

OH, IS THAT ALL?

(NOTE: HERE MOLECULAR WEIGHT REALLY MEANS THE MASS OF A BASIC PARTICLE OF THE SUBSTANCE EXPRESSED IN AMU. IN AN IONIC CRYSTAL LIKE NaCl, WE MEAN A BASIC COMPONENT OF THE CRYSTAL.

THE MOLE IS USED TO SCALE UP FROM ATOMIC DIMENSIONS TO METRIC WEIGHTS. TO BE PRECISE, A GRAM IS ABOUT 602,200,000,000,000,000,000,000 BIGGER THAN AN AMU. THAT IS, $1 g = 6.022 \times 10^{23}$ AMU.

ONE, TWO, THREE...

THIS THEN, IS THE **NUMBER OF PARTICLES IN A MOLE.** A MOLE OF **ANYTHING** HAS THIS MANY PARTICLES! 6.022×10^{23} IS CALLED **AVOGADRO'S NUMBER,** AFTER AMEDEO AVOGADRO, WHO FIRST SUGGESTED THAT EQUAL VOLUMES OF GAS HAVE EQUAL NUMBERS OF MOLECULES.

NOW SUPPOSE I START WITH 100 kg OF CLAY. HOW MANY KILOGRAMS OF POTTERY WILL I GET? WE START WITH THE BALANCED EQUATION:

$$3 Al_2Si_2O_5(OH)_4 (s) \xrightarrow{\Delta} Al_6Si_2O_{13}(s) + 4SiO_2(s) + 6H_2O(g)\uparrow$$

THE CLAY   THE POTTERY

THIS TAKES FOREVER WHEN YOUR BALANCE IS MADE OUT OF COCONUT SHELLS...

THEN WRITE A **MASS-BALANCE TABLE,** SHOWING THE NUMBER OF GRAMS OF EACH REACTANT AND PRODUCT:

REACTANTS	MOLAR WEIGHT	PRODUCTS	MOLAR WEIGHT
3 MOL $Al_2Si_2O_5(OH)_4$	3 × 258 = 774 g	1 MOL $Al_6Si_2O_{13}$	426 g
		4 MOL SiO_2	4 × 60 = 240 g
		6 MOL H_2O	6 × 18 = 108 g
TOTAL	774 g	TOTAL	774 g

THIS SAYS 774 g OF KAOLIN CLAY MAKES 426 + 240 = 666 g OF POTTERY.

SO 1 g KAOLIN MAKES (666/774) g = 0.86 g OF POTTERY

AND 100 kg MAKES (0.86)(100kg)(1000 g/kg) = 86,000 g = 86 kg.

WE CAN EQUALLY WELL WORK BACKWARD. IF WE WANT 100 kg OF POTTERY, HOW MUCH WET CLAY SHOULD WE MIX UP? (ANS: (100)(774/666) kg.)

PHEW!

More Reactions

WE'VE MADE VESSELS AND A STOVE. NOW LET'S COOK UP SOME **BUILDING MATERIALS.** WE HEAT LIMESTONE, CHALK, AND/OR SEASHELLS, WHICH ARE ALL MADE OF CALCIUM CARBONATE, $CaCO_3$. THE PRODUCT IS **QUICKLIME,** CaO.

$$CaCO_3\,(s) \xrightarrow{\Delta} CaO\,(s) + CO_2\,(g) \uparrow$$

BAKING CaO TOGETHER WITH POWDERED VOLCANIC ROCK MAKES **CEMENT.** ADD WATER, SAND, AND PEBBLES—**CONCRETE!** LET'S BUILD!

WE CAN EVEN PAINT OUR HOUSE. **WHITEWASH,** OR **SLAKED LIME,** $Ca(OH)_2$, COMBINES CaO AND H_2O:

$$CaO\,(s) + H_2O\,(l) \longrightarrow Ca(OH)_2\,(aq)$$

SLAKED LIME ALSO MAKES A GOOD PUTTY AND MORTAR... AND OVER TIME, WHITEWASH SLOWLY COMBINES WITH CO_2 FROM THE AIR AND HARDENS INTO A WHITE, STUCCO-LIKE MATERIAL:

$$Ca(OH)_2\,(s) + CO_2\,(g) \longrightarrow CaCO_3\,(s) + H_2O\,(g) \uparrow$$

LIMESTONE AGAIN!

NOW LET'S MAKE **SOAP,** SO WE CAN WASH UP.

FIRST BURN SEAWEED TO GET A WHITE, POWDERY MIXTURE OF Na_2CO_3 (SODA ASH) AND K_2CO_3 (POTASH). SEPARATE OUT THE SODA ASH (NEVER MIND HOW).

COMBINE SODA ASH WITH WHITEWASH TO MAKE THIS REACTION:

$$Ca(OH)_2(aq) + Na_2CO_3(aq)$$
$$\rightarrow$$
$$2NaOH(aq) + CaCO_3(s)\downarrow$$

A WHITE CLOUD OF $CaCO_3$ SETTLES TO THE BOTTOM. DECANT—CAREFULLY!—THE CLEAR NaOH SOLUTION. THIS IS **CAUSTIC LYE,** STRONG STUFF!

WE BOIL SOME **WILD BOAR FAT** WITH THE CAUSTIC LYE. THE FAT WILL NOT DISSOLVE IN WATER, BUT THE SODIUM IONS PUT A POLAR "TAIL" ON THE FAT MOLECULE, ALLOWING IT TO INTERACT WITH WATER IN A SOAPY WAY. WHAT'S THE REACTION?

WELL, THE **BOAR** IS **NOT** HAPPY!

FAT

+ 3 NaOH

GLYCEROL (A GOOD SKIN CONDITIONER)

A CRUDE SOAP

Redox Reactions

NOW LET'S MAKE SOME FLARES, SO WE CAN SIGNAL PASSING SHIPS. THIS WILL REQUIRE **EXPLOSIVE POWDER.** ITS INGREDIENTS ARE **CHARCOAL, SULFUR,** AND POTASSIUM NITRATE OR **SALTPETER,** KNO_3.

AND WE GET THIS FROM THE BOWELS OF THE—?

BATS!

WE ALREADY HAVE CHARCOAL... SULFUR WE SCRAPE UP IN ELEMENTAL FORM FROM THE NEARBY VOLCANO (IT'S THE YELLOW STUFF)... K IS IN POTASH, AND NITRATE WILL COME FROM $Ca(NO_3)_2$, WHICH WE FIND IN **BAT GUANO.**

YOU WANT ME TO **WHAT** IN **WHERE?**

BOIL THE GUANO IN WATER WITH POTASH AND GET A DOUBLE-DISPLACEMENT REACTION:

$$Ca(NO_3)_2 (aq) + K_2CO_3 (aq)$$
$$\longrightarrow$$
$$CaCO_3 (s)\downarrow + 2KNO_3 (aq)$$

THE CHALK SETTLES OUT OF SOLUTION.

WE CAREFULLY DECANT THE SOLUTION OF KNO_3.

LET THE WATER EVAPORATE AND WE ARE LEFT WITH A MASS OF NEEDLE-LIKE CRYSTALS OF KNO_3.

WHAT WILL THE REACTION PRODUCTS BE WHEN WE SET THIS STUFF OFF?

$$C + S + KNO_3 \longrightarrow ??$$

IT TURNS OUT THAT WE CAN MAKE A GOOD GUESS AT THE PRODUCTS BY FOLLOWING THE **ELECTRONS**.

EXPLOSIONS BELONG TO AN IMPORTANT CLASS OF REACTIONS INVOLVING THE **TRANSFER OF ELECTRONS** FROM ONE ATOM TO ANOTHER. SUCH REACTIONS ARE CALLED **OXIDATION-REDUCTION** REACTIONS, OR **REDOX** FOR SHORT.

WHY NOT "OXIDUCK?"

EXAMPLE: IN COMBUSTION,

$$C + O_2 \longrightarrow CO_2,$$

FOUR ELECTRONS MOVE FROM C TOWARD THE TWO O ATOMS. WE SAY C IS **OXIDIZED**. O, WHICH GAINS ELECTRONS, IS **REDUCED.** ANOTHER EXAMPLE IS RUSTING, OR **CORROSION:**

$$4Fe + 3O_2 \longrightarrow 2Fe_2O_3$$

Fe SHEDS ELECTRONS AND IS OXIDIZED; O GAINS THEM AND IS REDUCED.

NOTE: OXYGEN ITSELF NEED NOT BE INVOLVED! OXIDATION MEANS THE TRANSFER OF ELECTRONS TO **ANY** ATOM!

AS IN $H_2 + S \longrightarrow H_2S$, WHERE H IS OXIDIZED, AND SULFUR IS... UGH... REDUCED!

H_2S, ROTTEN EGG GAS

Oxidation Numbers

HOW MANY ELECTRONS DOES EACH ATOM GAIN OR LOSE?

THE **OXIDATION STATE** OR **OXIDATION NUMBER** OF AN ELEMENT IN A COMPOUND SHOWS ITS SURPLUS OR DEFICIT OF ELECTRONS. THAT IS, THE OXIDATION NUMBER IS THE **NET CHARGE ON THE ATOM.*** FOR INSTANCE, IN CaO, Ca HAS THE OXIDATION NUMBER +2—IT GIVES AWAY TWO ELECTRONS—AND O'S OXIDATION NUMBER IS −2, BECAUSE IT ACCEPTS TWO.

1) THE OXIDATION NUMBER OF AN ELEMENT IN ELEMENTAL FORM IS ZERO.

2) SOME ELEMENTS HAVE THE SAME OXIDATION NUMBER IN ALMOST ALL THEIR COMPOUNDS:

- H: +1 (EXCEPT IN METAL HYDRIDES LIKE NaH, WHERE IT'S −1)
- ALKALI METALS Li, Na, K, ETC.: +1
- GROUP 2 METALS Be, Mg, ETC.: +2
- FLUORINE: −1
- OXYGEN: ALMOST ALWAYS −2

3) IN A NEUTRAL COMPOUND, THE OXIDATION NUMBERS ADD UP TO ZERO.

4) IN A POLYATOMIC ION, THE OXIDATION NUMBERS ADD UP TO THE CHARGE ON THE ION.

---

*OR WHAT IT WOULD BE, IF THE BOND WERE FULLY IONIC. IN ASSIGNING OXIDATION NUMBERS, WE PRETEND THAT THE ELECTRONS ARE COMPLETELY TRANSFERRED FROM ONE ATOM TO ANOTHER, EVEN THOUGH IN REALITY THEY MAY BE ONLY UNEQUALLY SHARED.

AN ATOM'S OXIDATION NUMBER DEPENDS ON THE OTHER ATOMS AROUND IT. FOR INSTANCE, IN HCl, CHLORINE ACQUIRES ONE ELECTRON (FOR AN OXIDATION STATE OF −1) BECAUSE Cl IS MORE ELECTRONEGATIVE (EN = 3.0) THAN HYDROGEN (EN = 2.1),

BUT IN THE **PERCHLORATE** ION, ClO_4^-, CHLORINE HAS AN OXIDATION NUMBER OF +7. ALL ITS VALENCE ELECTRONS GO TO OXYGEN, WHICH IS EVEN MORE ELECTRO-NEGATIVE (EN = 3.5) THAN CHLORINE.

HERE ARE SOME ELEMENTS AND THEIR COMMON OXIDATION NUMBERS. THE BIGGER THE PLUS, THE MORE OXIDIZED.

	MOST REDUCED	INTERMEDIATE	MOST OXIDIZED
H	NiH_2 (−1)	H_2 (0)	H_2O, OH^- (+1)
C	CH_4 (−4)	C (0)	CO_2, CO_3^{2-} (+4)
O	H_2O, CO_2, CaO, ETC. (−2)	H_2O_2 (−1) (HYDROGEN PEROXIDE)	O_2 (0)
N	NH_3 (−3)	N_2 (0), N_2O (+1), NO (+2)	NO_3^- (+5)
S	H_2S, K_2S (−2)	S (0), SO_2 (+4)	SO_3, SO_4^{2-} (+6)
Fe	Fe (0)	FeO (+2)	Fe_2O_3 (+3)
Cl	HCl (−1)	Cl_2 (0)	ClO_4^- (+7)

→ OXIDATION

← REDUCTION

IN REDOX REACTIONS, SOME SUBSTANCES—**REDUCING AGENTS** OR **REDUCTANTS**—DONATE ELECTRONS, AND OTHERS—**OXIDIZING AGENTS, OR OXIDANTS**—GAIN THEM.

WAIT—THE OXIDIZING AGENT IS REDUCED AND THE REDUCING AGENT IS OXIDIZED?

RIGHT!

SO THE ONE THAT'S REDUCED **GAINS** ELECTRONS? SOUNDS WRONG...

ITS **CHARGE** IS REDUCED, ISN'T IT?

REDUCING AGENT

OXIDIZING AGENT

REDUCED

OXIDIZED

GOING BACK TO THE EXPLOSIVE BLACK POWDER, WHAT ARE THE MOST LIKELY OXIDIZING AND REDUCING AGENTS? LET'S IGNORE THE SULFUR FOR THE TIME BEING AND CONCENTRATE ON THE CARBON AND SALTPETER:

$$C + KNO_3 \longrightarrow ?$$

OF THOSE FOUR ELEMENTS, WE CAN ELIMINATE K AND O, BECAUSE THEY ARE ALREADY FULLY OXIDIZED (K AT +1) AND REDUCED (O AT -2) RESPECTIVELY. IT IS VERY HARD TO OXIDIZE O^{2-} OR REDUCE K^+! BUT C (0) CAN BE OXIDIZED TO +4 AS EITHER CO_2 OR CO_3^{2-}, AND N (+5) CAN BE REDUCED TO 0 AS N_2. SO WE EXPECT SOMETHING LIKE THIS BEFORE BALANCING:

$$C(s) + KNO_3(s) \longrightarrow CO_2(g)\uparrow + N_2(g)\uparrow + K_2CO_3(s)$$

WE CAN BALANCE THIS BY FOLLOWING THE ELECTRONS: EACH MOL OF C GIVES UP 4 MOL ELECTRONS, AND EACH MOL OF N ACCEPTS 5. THIS BALANCES IF **20** MOL ELECTRONS MOVE FROM 5C TO 4N. (WE GET THE OTHER COEFFICIENTS BY BALANCING K AND O.)

$$5C(s) + 4KNO_3(s) \longrightarrow 3CO_2(g)\uparrow + 2N_2(g)\uparrow + 2K_2CO_3(s)$$

THIS REACTION WILL ACTUALLY PRODUCE A PRETTY GOOD FIZZ, BUT CENTURIES OF EXPERIMENT HAVE SHOWN THAT ADDING SULFUR MAKES A MUCH BIGGER POP.

ELEMENTAL S (0), REDUCES EASILY TO -2 IN K_2S. IN FACT, CHEMISTS NOW KNOW THAT FORMING K_2S IS "EASIER" THAN FORMING K_2CO_3. DOING SO CONSUMES LESS ENERGY—AND LEAVES MORE ENERGY TO POWER THE BANG. SO WE EXPECT SOMETHING LIKE:

$$C(s) + KNO_3(s) + S(s) \longrightarrow CO_2(g)\uparrow + N_2(g)\uparrow + K_2S(s)$$

EACH C LOSES **4** ELECTRONS   EACH N GAINS **5** ELECTRONS   EACH S GAINS **2** ELECTRONS

THIS BALANCES WHEN 3 MOLS C GIVE UP 12 MOLS ELECTRONS, OF WHICH 10 MOLS ELECTRONS GO TO 2 MOLS N AND 2 MOLS ELECTRONS GO TO ONE MOL S:

$$3C(s) + 2KNO_3(s) + S(s) \longrightarrow 3CO_2(g)\uparrow + N_2(g)\uparrow + K_2S(s) + \textbf{BANG!}$$

NOW WE CAN MAKE A FORMULA FOR BLACK POWDER. WE START WITH THE MASS-BALANCE TABLE:

REACTANTS	MOLAR WEIGHT	PRODUCTS	MOLAR WEIGHT
3 mol C	$3 \times 12 = 36$ g	3 mol CO_2	$3 \times 44 = 132$ g
2 mol KNO_3	$2 \times 101 = 202$ g	1 mol N_2	28 g
1 mol S	32 g	1 mol K_2S	110 g
TOTAL	270 g	TOTAL	270 g

FOR ONE GRAM OF POWDER, WE NEED $(36/270)$ g $= 0.13$ g C, $(202/270)$ g $= 0.75$ g KNO_3, AND $(32/270)$ g $= 0.12$ g S. MULTIPLY BY 100 TO SEE WHAT WE NEED TO MAKE 100 g OF POWDER:

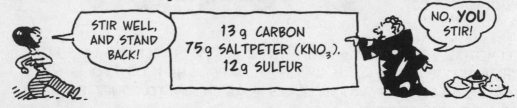

STIR WELL, AND STAND BACK!

13 g CARBON
75 g SALTPETER (KNO_3).
12 g SULFUR

NO, **YOU** STIR!

NOT BAD! A CLASSIC GUNPOWDER RECIPE CALLS FOR 10g SULFUR, 15g CARBON, AND 75g SALTPETER. THE DIFFERENCE FROM OUR RESULT IS DUE TO TRACES OF OTHER REACTION PRODUCTS THAT WE NEGLECTED. THE REAL RECIPE IS A PRODUCT OF TRIAL AND ERROR.

YOU STIR!

NO, YOU!

YOU!

NOW WE MIX SOME OF THIS STUFF UP...

YOU!

OH, ALL RIGHT...

IF YOU TRY THIS AT HOME **(NOT RECOMMENDED** IN THE FIRST PLACE), ALWAYS BE SURE TO GRIND THE INGREDIENTS **SEPARATELY—** UNLESS YOU WANT TO BLOW OFF YOUR FINGERS, OR EVEN YOUR WHOLE HAND.

WE PACK OUR POWDER INTO BAMBOO TUBES, AND—SAY, HERE COMES A SHIP! LIGHT THE FUSE!

AHOY!

Chapter 5
Heat of Reaction

IN THE LAST CHAPTER, WE LOOKED AT CHEMICAL REACTIONS AS TRANSFERS OF MATTER. WE KEPT A CAREFUL ACCOUNTING OF ATOMS AS THEY REARRANGED THEMSELVES.

NOW WE LOOK AT REACTIONS ANOTHER WAY: AS TRANSFERS OF **ENERGY**.

ENERGY? WHAT ENERGY?

PHYSICISTS DEFINE ENERGY MECHANICALLY, AS THE ABILITY TO DO **WORK.*** WORK IS WHAT HAPPENS WHEN A FORCE OPERATES ON AN OBJECT OVER A DISTANCE: WORK = FORCE X DISTANCE. THE METRIC UNIT OF ENERGY IS THE NEWTON-METER, OR **JOULE.**

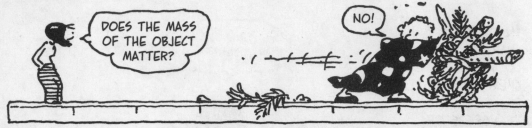

DOES THE MASS OF THE OBJECT MATTER?

NO!

1 JOULE = WORK DONE BY A FORCE OF ONE NEWTON OPERATING OVER A DISTANCE OF ONE METER.

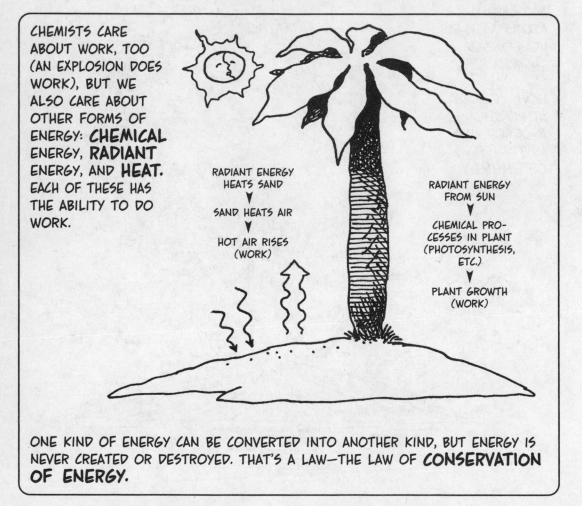

CHEMISTS CARE ABOUT WORK, TOO (AN EXPLOSION DOES WORK), BUT WE ALSO CARE ABOUT OTHER FORMS OF ENERGY: **CHEMICAL** ENERGY, **RADIANT** ENERGY, AND **HEAT.** EACH OF THESE HAS THE ABILITY TO DO WORK.

RADIANT ENERGY HEATS SAND
▼
SAND HEATS AIR
▼
HOT AIR RISES (WORK)

RADIANT ENERGY FROM SUN
▼
CHEMICAL PROCESSES IN PLANT (PHOTOSYNTHESIS, ETC.)
▼
PLANT GROWTH (WORK)

ONE KIND OF ENERGY CAN BE CONVERTED INTO ANOTHER KIND, BUT ENERGY IS NEVER CREATED OR DESTROYED. THAT'S A LAW—THE LAW OF **CONSERVATION OF ENERGY.**

*NOT TO BE CONFUSED WITH USEFUL WORK.

LET'S EXAMINE MECHANICAL ENERGY MORE CLOSELY. IF I PUSH THIS COCONUT, IT MOVES... AND THE LONGER AND/OR HARDER I PUSH, THE FASTER IT GOES. (THIS IS CLEARER IN OUTER SPACE, AWAY FROM FRICTION AND GRAVITY.) BY DOING WORK ON THE COCONUT, I ADD ENERGY TO IT: **KINETIC ENERGY (K.E.),** THE ENERGY OF MOTION.

$$K.E. = \frac{1}{2}mv^2$$

BACK ON EARTH, I PUSH THE COCONUT AGAIN, BUT IN AN UPWARD DIRECTION. THE COCONUT FLIES UP, BUT IT SLOWS UNDER THE PULL OF GRAVITY. EVENTUALLY IT STOPS AND BEGINS TO FALL. WHAT BECAME OF THE ENERGY I ADDED??

STATIONARY, NO K.E., HIGH P.E.

LOW SPEED, SOME K.E., SOME P.E.

HIGH SPEED, HIGH K.E.

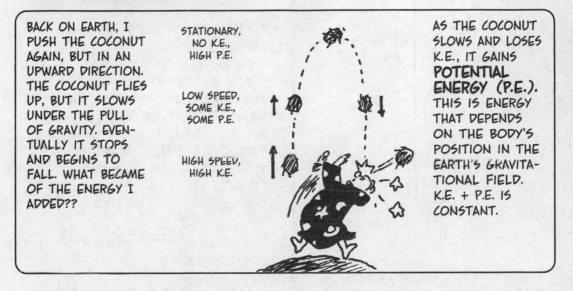

AS THE COCONUT SLOWS AND LOSES K.E., IT GAINS **POTENTIAL ENERGY (P.E.).** THIS IS ENERGY THAT DEPENDS ON THE BODY'S POSITION IN THE EARTH'S GRAVITATIONAL FIELD. K.E. + P.E. IS CONSTANT.

IT TURNS OUT THAT **ALL** FORMS OF ENERGY CAN BE UNDERSTOOD IN TERMS OF KINETIC AND POTENTIAL ENERGY. RADIANT ENERGY, FOR INSTANCE, IS THE K.E. OF MOVING PHOTONS, OR LIGHT PARTICLES.* THERE IS POTENTIAL ENERGY STORED IN CHEMICAL BONDS. AND HEAT IS... HEAT IS... WHAT **IS** HEAT, ANYWAY?

IT'S HARD TO DESCRIBE WITHOUT SCREAMING.

*THE "LIGHT" NEED NOT BE VISIBLE. MOVING PHOTONS CONVEY THE ENERGY OF ALL ELECTROMAGNETIC RADIATION, FROM X-RAYS TO RADIO WAVES.

HEAT, WE KNOW, HAS SOMETHING TO DO WITH **TEMPERATURE,** AND TEMPERATURE IS FAMILIAR ENOUGH. WE EVEN KNOW HOW TO MEASURE IT, WITH A THERMOMETER.

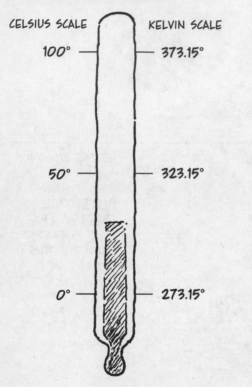

CELSIUS SCALE	KELVIN SCALE
100°	373.15°
50°	323.15°
0°	273.15°

THE UNITS ARE **DEGREES CELSIUS** (°C). THE CELSIUS SCALE SETS:

0°C = MELTING POINT OF WATER
100°C = BOILING POINT OF WATER

THE **KELVIN** SCALE HAS DEGREES THE SAME SIZE AS CELSIUS, BUT STARTS LOWER:

0°K = ABSOLUTE ZERO, WHERE ALL MOLECULAR AND ATOMIC MOTION STOPS = −273.15°C.

°C = °K − 273.15

COLLOQUIALLY, WE SAY SOMETHING IS HOT WHEN WE REALLY MEAN IT HAS A HIGH TEMPERATURE. A CHEMIST WOULD NEVER SAY THIS! **HEAT AND TEMPERATURE ARE NOT THE SAME.**

BUT NOT EXACTLY DIFFERENT EITHER...

TO ILLUSTRATE THE DIFFERENCE, SUPPOSE WE COOK TWO COCONUTS, RAISING THEIR TEMPERATURE BY 75°C (FROM 25° TO 100°, SAY). THEN THE TWO COCONUTS TOGETHER HAVE THE **SAME TEMPERATURE CHANGE** AS ONE COCONUT, BUT THEY ABSORB **TWICE AS MUCH HEAT,** BECAUSE THEY CONTAIN TWICE AS MUCH MATTER TO HEAT UP.

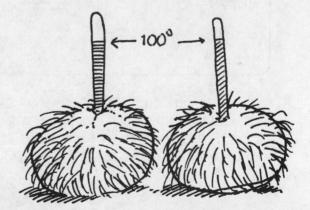

SAME TEMPERATURE CHANGE
DOUBLE THE HEAT CHANGE

WHAT, THEN, IS THE RELATIONSHIP BETWEEN TEMPERATURE AND HEAT?

TO BEGIN WITH, WHEREVER WE LOOK, HEAT TRANSFERS ARE ASSOCIATED WITH **TEMPERATURE DIFFERENCES.** WE KNOW FROM EXPERIENCE THAT HEAT FLOWS FROM HOT TO COLD.

IT'S AN ENERGY CHANGE THAT INVOLVES NO VISIBLE WORK OR MOVEMENT!

YOW! WHO SAYS?

THAT IS, WHEN A HIGHER-TEMPERATURE OBJECT MEETS A LOWER-TEMPERATURE OBJECT, ENERGY FLOWS FROM THE WARMER ONE TO THE COOLER ONE UNTIL THEIR TEMPERATURES EQUALIZE. AN EXAMPLE IS WHEN WE IMMERSE SOMETHING COOL IN HOT WATER. (ASSUME THAT THE "SOMETHING" DOESN'T MELT.)

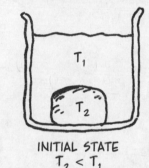

INITIAL STATE
$T_2 < T_1$

HEAT FLOW TAKES PLACE

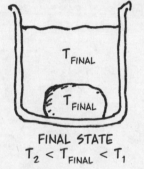

FINAL STATE
$T_2 < T_{FINAL} < T_1$

(FINAL TEMPERATURES ARE EQUAL, AND BETWEEN THE ORIGINAL EXTREMES)

THE AMOUNT OF ENERGY TRANSFERRED IS THE HEAT: **HEAT IS THE ENERGY CHANGE ASSOCIATED WITH A DIFFERENCE IN TEMPERATURE.**

IT WORKS FOR HOT THINGS IN COLD WATER TOO... AHH...

Internal Energy

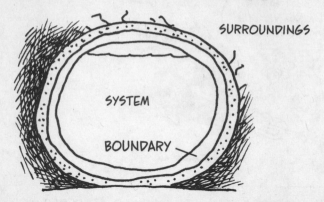

SURROUNDINGS

SYSTEM

BOUNDARY

WHERE DOES HEAT ENERGY GO? TO ANSWER THIS QUESTION, CONSIDER THIS COCONUT, WHICH REALLY STANDS FOR ANY CHEMICAL SYSTEM WITH A DEFINITE BOUNDARY BETWEEN ITSELF AND ITS SURROUNDINGS.

AT CLOSE RANGE, THE COCONUT SEETHES WITH ENERGY. ALL ITS MOLECULES ARE JIGGLING RANDOMLY, SO THEY HAVE KINETIC ENERGY. THEY ALSO HAVE POTENTIAL ENERGY: ELECTRIC ATTRACTIONS AND REPULSIONS ACCELERATE AND DECELERATE PARTICLES, ANALOGOUS TO THE WAY GRAVITY WORKS ON A THROWN OBJECT.

NOT SO FAST...

A SYSTEM'S **INTERNAL ENERGY** IS THE TOTAL KINETIC AND POTENTIAL ENERGY OF ALL ITS PARTICLES.

AN ENERGY YOU CAN'T SEE...

BUT YOU CAN FEEL!

A SYSTEM'S **TEMPERATURE** IS A MEASURE OF THE AVERAGE TRANSLATIONAL KINETIC ENERGY* OF ALL ITS PARTICLES, I.E., HOW FAST THEY FLY OR WIGGLE.

THIS MAKES SENSE, GIVEN WHAT WE KNOW ABOUT TEMPERATURE. A HIGHER-T SYSTEM RAISES THE TEMPERATURE OF A LOWER-T SYSTEM BECAUSE HIGHER-ENERGY PARTICLES TRANSFER ENERGY TO LOWER-ENERGY ONES.

SOMETHING LIKE BILLIARD BALLS!

THIS IS A BIT MORE COMPLICATED THAN IT SOUNDS. IN GASES, T MEASURES HOW ENERGETICALLY MOLECULES FLY AROUND, BUT IN METALS, T ALSO INCLUDES THE ENERGY OF MOVING ELECTRONS... IN CRYSTALS, WIGGLING IONS HAVE P.E. AS WELL AS K.E., BECAUSE PARTICLES PULL AGAINST EACH OTHER... AND MOLECULES (OR PARTS OF MOLECULES) CAN ROTATE OR VIBRATE INTERNALLY. EVERY SUBSTANCE IS DIFFERENT!

WHEN HEAT IS ADDED AND INTERNAL ENERGY RISES, SOME OF THE ADDED ENERGY DOES NOT CONTRIBUTE TO A RISE IN TEMPERATURE, BUT RATHER IS ABSORBED AS P.E., ROTATION, OR INTERNAL VIBRATION.

IT'S LIKE BILLIARDS WITH WEIRD-SHAPED, SPRINGY BALLS!

THAT IS:

Different chemicals have different temperature responses to heat.

GAH! THESE HANDS!

*TRANSLATIONAL ENERGY IS ENERGY ASSOCIATED WITH PARTICLES MOVING THROUGH SPACE. THE ENERGY OF SPINNING AND INTERNAL VIBRATION IS NOT INCLUDED.

Heat Capacity

THE **HEAT CAPACITY** OF A SUBSTANCE IS THE ENERGY INPUT REQUIRED TO RAISE ITS TEMPERATURE BY 1°C. WE CAN SPEAK OF HEAT CAPACITY PER GRAM ("SPECIFIC HEAT") OR PER MOLE ("MOLAR HEAT CAPACITY").

IN OTHER WORDS, IT'S THE... UM... CAPACITY... OF THE SUBSTANCE TO SOAK UP... ER... HEAT...

YOU'RE SO ELOQUENT!

JAMES PRESCOTT **JOULE** (1818–1889) MEASURED THE HEAT CAPACITY OF WATER. HE ATTACHED A FALLING WEIGHT TO A PADDLE WHEEL IMMERSED IN WATER. BY MEASURING THE SLIGHT RISE IN TEMPERATURE OF THE WATER,* JOULE FOUND THE WORK EQUIVALENT OF A TEMPERATURE CHANGE. RESULT:

WATER'S HEAT CAPACITY PER GRAM OR **SPECIFIC HEAT** IS

4.184 Joules/g°C

EXAMPLE: TO RAISE THE TEMPERATURE OF 5g OF WATER BY 7°C REQUIRES AN ADDED ENERGY OF

5 X 7 X 4.184
= 146 JOULES.

*YOU CAN RAISE TEMPERATURE BY DOING WORK ON AN OBJECT. FOR INSTANCE, WHEN YOU HAMMER A NAIL, THE NAIL HEAD WARMS UP.

Heat change = Mass × ΔT × Specific heat

ΔT?

CHANGE IN TEMPERATURE.

FROM THAT SINGLE FORMULA AND WATER'S SPECIFIC HEAT, WE CAN FIND ALL OTHER SPECIFIC HEATS! LET'S START WITH COPPER. IMMERSE 2 kg COPPER AT 25°C IN 5 kg WATER AT 30°C. LET THE TEMPERATURE STABILIZE. CHECK THE THERMOMETER. IT READS 29.83°C. THE WATER BARELY CHANGED TEMPERATURE, BUT THE COPPER REALLY HEATED UP!

5 kg AT 30°

2 kg AT 25°

29.83°C

29.83°C

THE TEMPERATURE CHANGES (ΔT) ARE

$$\Delta T_{WATER} = -0.17°$$
$$\Delta T_{COPPER} = 4.83°$$

WE CAN IMMEDIATELY CALCULATE WATER'S HEAT LOSS. (HEAT CHANGES ARE DENOTED BY THE LETTER q):

$$q_{WATER} = (5000g)(-0.17°C)(4.18 \text{ J/g°C})$$
$$= -3553 \text{ Joules}$$

THE MINUS SIGN MEANS THAT THE WATER GAVE UP ENERGY.

BUT THE WATER'S LOSS IS PRECISELY COPPER'S GAIN (ASSUMING NO HEAT LEAKS OUT OF THE VESSEL). THAT IS,

$$q_{COPPER} = 3553 \text{ Joules.}$$

SINCE THERE WERE 2000g OF COPPER, THE FORMULA SAYS:

$$3553 \text{ J} = (2000g)(4.83°)C_{Cu}$$

(C_{Cu} = COPPER'S SPECIFIC HEAT)

SOLVING FOR C_{Cu},

$$C_{Cu} = \frac{3553 \text{ J}}{(2000 g)(4.83°)} = 0.37 \text{ J/g°C}$$

AMAZINGLY, COPPER'S SPECIFIC HEAT IS LESS THAN **ONE-TENTH** THAT OF WATER. WATER CAN SOAK UP HEAT WITH LITTLE RISE IN TEMPERATURE, WHILE COPPER'S TEMPERATURE RISES ALMOST EFFORTLESSLY.

THE DIFFERENCE COMES FROM THEIR INFERNAL STRUCTURES!

THAT'S "INTERNAL"! IN**TER**NAL!

LIQUID WATER HAS MANY HYDROGEN BONDS BETWEEN ITS MOLECULES (SEE CHAPTER 3). THESE BONDS MAKE IT HARD TO GET A WATER MOLECULE MOVING! ADDED HEAT LARGELY GOES INTO THE P.E. ASSOCIATED WITH THESE ATTRACTIONS.

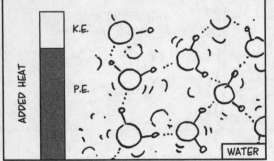

K.E.

P.E.

ADDED HEAT

WATER

COPPER, ON THE OTHER HAND, HAS A "SEA" OF HIGHLY MOBILE ELECTRONS. ADDED ENERGY SIMPLY MAKES THEM FLY AROUND FASTER. THAT IS, HEAT ALMOST ALL GOES INTO K.E., AND TEMPERATURE RISES ACCORDINGLY.

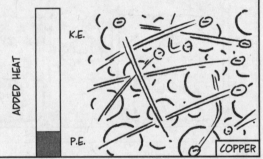

K.E.

P.E.

ADDED HEAT

COPPER

THIS EXPLAINS WHY WATER IS USED AS A COOLANT IN MACHINERY, FROM CAR ENGINES TO NUCLEAR REACTORS. THE HEAT TRANSFER FROM HOT METAL TO COOL WATER DROPS THE METAL'S TEMPERATURE DRAMATICALLY, WHILE RAISING WATER'S RELATIVELY LITTLE.

SEE? I TOLD YOU I MEANT INFERNAL...

MANY OTHER SPECIFIC HEATS CAN BE FOUND THE SAME WAY. IF WE REPLACE COPPER WITH IRON IN THE EXPERIMENT (SAME TEMPERATURES, SAME MASSES), WE FIND

$$\Delta T_{WATER} = -0.206°$$

$$\Delta T_{IRON} = 4.794°$$

FROM THE EXACT SAME COMPUTATION AS BEFORE, WE FIND

$$C_{IRON} = 0.45 \; J/g°C$$

ALSO VERY LOW.

NOW MEASURE IRON AGAINST ETHANOL, OR GRAIN ALCOHOL. ASSUME THE SAME MASSES AND A 5° TEMPERATURE DIFFERENCE AT THE START.

$$\Delta T_{ETHANOL} = -0.36°$$

$$\Delta T_{IRON} = 4.65°$$

AND WE CALCULATE AS BEFORE:

$$C_{ETHANOL} = 2.4 \; J/g°C$$

CLOSER TO WATER.

WE CAN CONTINUE MEASURING ONE THING AGAINST ANOTHER UNTIL WE "BOOTSTRAP" A WHOLE TABLE OF SPECIFIC HEATS.

SUBSTANCE	SPECIFIC HEAT $(J/g°C)$
MERCURY, Hg	0.14
COPPER, Cu	0.37
IRON, Fe	0.45
C (GRAPHITE)	0.68
SIMPLE MOLECULES	
ICE, H_2O (s)	2.0
WATER VAPOR, H_2O (g)	2.1
ANTIFREEZE, (CH_2OHCH_2OH)	2.4
ETHANOL, (CH_3CH_2OH)	2.4
LIQUID WATER, $H_2O(l)$	4.2
AMMONIA, $NH_3(l)$	4.7
COMPLEX MATERIALS	
BRASS	0.38
GRANITE	0.79
GLASS	0.8
CONCRETE	0.9
WOOD	1.8

Note that antifreeze is a less effective coolant than water, but it has the advantages of having a lower freezing point and being less corrosive to engine parts.

YES, O ALL-KNOWING HAND!

Calorimetry

THE POINT OF ALL THESE PRELIMINARIES IS TO FIND THE **HEAT CHANGES OF CHEMICAL REACTIONS:** HOW MUCH ENERGY IS RELEASED OR ABSORBED AS HEAT WHEN A REACTION TAKES PLACE. WE ARE NOW IN A POSITION TO MEASURE THIS.

THE METHOD IS SIMILAR TO THE WAY WE FOUND SPECIFIC HEATS: RUN THE REACTION IN A VESSEL OF KNOWN HEAT CAPACITY C AND MEASURE THE CHANGE IN TEMPERATURE. SINCE THE VESSEL ABSORBS WHAT THE REACTION GIVES OFF—OR VICE VERSA—THE HEAT CHANGE q OF THE REACTION IS $-q_{VESSEL} = -C\Delta T$.

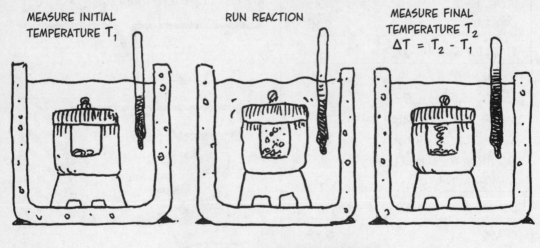

MEASURE INITIAL TEMPERATURE T_1

RUN REACTION

MEASURE FINAL TEMPERATURE T_2
$\Delta T = T_2 - T_1$

$$q = -C\Delta T$$

THE REACTION VESSEL AND ITS SURROUNDING PARAPHERNALIA TOGETHER ARE CALLED A **BOMB CALORIMETER.** THE REACTION CHAMBER, OR "BOMB," IS USUALLY IMMERSED IN WATER, WHICH CAN BE STIRRED TO DISTRIBUTE THE HEAT. A THERMOMETER COMPLETES THE APPARATUS.

Example

COMBUSTION OF **OCTANE** C_8H_{18}, A COMPONENT OF GASOLINE:

$$2C_8H_{18}(l) + 25O_2(g) \longrightarrow 16CO_2(g) + 18H_2O(g)$$

TO MEASURE THE HEAT GIVEN OFF, WE NEED A STRONG, HEAVY BOMB TO WITH-STAND THE HIGH TEMPERATURE AND PRESSURE GENERATED. A THICK-WALLED STEEL CONTAINER OUGHT TO DO... LET'S SUPPOSE ITS HEAT CAPACITY IS 15,000 J/°C. WE IMMERSE IT IN **2.5 L** OF WATER, WHICH HAS A MASS OF 2500 g.

THE WATER'S HEAT CAPACITY IS

$$(2500g)(4.184 J/g°C) = 10,460 \ J/°C.$$

SO THE CALORIMETER'S TOTAL HEAT CAPACITY IS

$$10,460 + 15,000 = 25,460 \ J/°C.$$

SUPPOSE T_1, THE INITIAL TEMPERATURE OF THE CALORIMETER, IS 25°.

ONCE YOU KNOW SPECIFIC HEATS, IT'S ALL TEMPERATURE-TAKING!

WE DROP ONE GRAM OF OCTANE INTO THE BOMB... IGNITE IT WITH A SPARK... IT BURNS... THE HEAT SPREADS THROUGHOUT THE CALORIMETER... WE AGAIN CONSULT THE THERMOMETER, AND FIND $T_2 = 26.88°$. THEN

$$\Delta T = T_2 - T_1 = 1.88°$$

THE MAGIC FORMULA IS

$$q = -C_{CALORIMETER}(\Delta T)$$

WE PLUG IN AND FIND

$$q = -(25,460 \ J/°C)(1.88°C) = -47,800 \ J$$
$$= -47.8 \ kJ$$

AND WE CONCLUDE THAT OCTANE RELEASES 47.8 kJ/g OF HEAT WHEN BURNED.

IS THAT A LOT, 47.8 kJ/g?

A FORCE OF ONE NEWTON PUSHING FOR 47.8 KILOMETERS? **YES**, IT'S A LOT!!

Enthalpy

THE BOMB CALORIMETER IS GREAT, WONDERFUL, FANTASTIC, BUT A BIT UNREALISTIC, BECAUSE THE REACTION VESSEL IS SEALED. SOME REACTIONS IN THE BOMB MAY PRODUCE HIGH PRESSURES, WHICH CAN AFFECT TEMPERATURE.

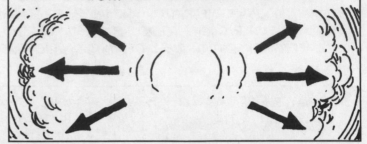

FOR EXAMPLE, AN EXPLOSION IN THE OPEN AIR GIVES OFF GASES THAT EXPAND RAPIDLY AND PUSH THE SURROUNDING AIR OUTWARD. IN OTHER WORDS, THE GASES DO **WORK** ON THE SURROUNDINGS.

IN THAT CASE, THE ENERGY CHANGE ΔE OF THE REACTION HAS TWO COMPONENTS, **WORK** AND **HEAT**:

$$\Delta E = \Delta H + \text{WORK}$$

PUSHING AIR OUT OF THE WAY COOLS THE REACTION PRODUCTS!

HEAT CHANGE

WORK

TOTAL ENERGY CHANGE

ΔH HERE MEANS THE HEAT CHANGE WHEN THE REACTION IS RUN OUTDOORS.

IN THE BOMB CALORIMETER, THE GASES DO NO WORK, BECAUSE THE EXPLOSION IS CONFINED IN A FIXED VOLUME. **ALL** THE ENERGY IS RELEASED AS HEAT.

$$\Delta E = q$$

THEREFORE

$$q = \Delta H + \text{WORK}$$

SO

$$q > \Delta H$$

THE HEAT CHANGE IN THE BOMB IS GREATER THAN THAT IN THE OUTSIDE WORLD.

FROM NOW ON, WE TREAT REACTIONS AS IF THEY TAKE PLACE "OUTDOORS"—MEANING AT **CONSTANT PRESSURE.** IN THAT CASE, THE HEAT RELEASED OR ABSORBED IS CALLED THE **ENTHALPY CHANGE,** AND WRITTEN ΔH.

POSSIBLY THE UGLIEST WORD IN ALL CHEMISTRY!

TO MEASURE ENTHALPY CHANGE, WE USE A CALORIMETER THAT MAINTAINS CONSTANT PRESSURE. THEN THE PROCEDURE IS THE SAME AS WITH A BOMB CALORIMETER: MEASURE INITIAL AND FINAL TEMPERATURES T_1 AND T_2, THEN MULTIPLY $T_2 - T_1$ TIMES THE HEAT CAPACITY OF THE CALORIMETER.

Example

EXPLOSION OF BLACK POWDER (HERE WE GIVE A MORE REALISTIC EQUATION THAN PREVIOUSLY):

$$4KNO_3(s) + 7C(s) + S(s) \longrightarrow 3CO_2\uparrow + 3CO\uparrow + 2N_2\uparrow + K_2CO_3(s) + K_2S(s)$$

SUPPOSE OUR CALORIMETER HAS A KNOWN HEAT CAPACITY OF 337.6 kJ/°C. WE START WITH 500g OF POWDER. THE TEMPERATURE CHANGE ΔT IS FOUND TO BE 4.78°C, AND WE COMPUTE

$$\Delta H = -(337.6 \text{ kJ/°C})(4.78°C)$$
$$= -1614 \text{ kJ}$$

FROM THIS WE CAN FIND THE ENTHALPY CHANGE PER GRAM, $\Delta H/g$.

$$\Delta H/\text{gram} = \frac{-1614}{500} = -3.23 \text{ kJ/g}$$

Example

HERE IS A REACTION THAT ABSORBS HEAT:

$$CaCO_3(s) \xrightarrow{\Delta} CaO(s) + CO_2\uparrow$$

WE START WITH THE CALORIMETER HOT ENOUGH TO DRIVE THE REACTION. AT THE END, THE CALORIMETER IS **COOLER** THAN AT THE BEGINNING. IF WE START WITH ONE MOLE OF $CaCO_3$, WE FIND THAT

$$\Delta T = -0.53°C$$

SO

$$\Delta H = -(337.6 \text{ kJ/°C})(-0.53°C)$$
$$= 179 \text{ kJ/mol}$$

REACTIONS THAT RELEASE HEAT ($\Delta H < 0$) ARE CALLED **EXOTHERMIC**. REACTIONS THAT ABSORB HEAT FROM THE SURROUNDINGS ($\Delta H > 0$) ARE CALLED **ENDOTHERMIC**.

Heats of Formation

GREAT! NOW WE CAN MEASURE ΔH FOR JUST ABOUT ANY REACTION! TOO BAD THERE ARE SO MANY REACTIONS... THIS COULD TAKE A WHILE... LUCKILY, INGENIOUS (OR LAZY) CHEMISTS HAVE THOUGHT UP A **SHORT CUT:** INSTEAD OF MEASURING ENTHALPY CHANGES, WE CAN **CALCULATE** THEM.

SCIENCE IS A KIND OF SYSTEMATIC LAZINESS...

THE BASIC CONCEPT IS CALLED **ENTHALPY OF FORMATION,** WRITTEN ΔH_f: THE ENTHALPY CHANGE THAT OCCURS WHEN A MOLE OF SUBSTANCE IS FORMED FROM ITS CONSTITUENT ELEMENTS. FOR INSTANCE, WHEN A MOLE OF LIQUID WATER IS FORMED FROM HYDROGEN AND OXYGEN, OUR CALORIMETER MEASURES

$$H_2(g) + \tfrac{1}{2}O_2(g) \longrightarrow H_2O(l) \quad \Delta H_f = \Delta H = -285.8 \text{ kJ/mole}$$

EACH SUBSTANCE HAS A HEAT OF FORMATION, WHICH CAN EITHER BE MEASURED OR INFERRED. EVERY ELEMENT IN ITS MOST STABLE FORM (SUCH AS C, O_2 OR S) HAS $\Delta H_f = 0$.

SUBSTANCE	ΔH_f, kJ/mol
$CO(g)$	−110.5
$CO_2(g)$	−393.8
$CaCO_3(s)$	−1207.6
$CaO(s)$	−635.0
$H_2O(l)$	−285.8
$H_2O(g)$	−241.8
$S(s)$	0
$KNO_3(s)$	−494.0
$K_2CO_3(s)$	−1151.0
$C_3H_5(NO_3)_3(l)$	−364.0
$N_2(g)$	0
$O_2(g)$	0

WHY TWO ENTRIES FOR WATER, ONE FOR LIQUID AND ONE FOR GAS?

TO BE EXPLAINED NEXT CHAPTER...

HOW DO WE USE HEATS OF FORMATION? HERE'S THE IDEA. IMAGINE ANY REACTION: REACTANTS → PRODUCTS. LET'S IMAGINE IT AS **TWO SUCCESSIVE REACTIONS:** REACTANTS → CONSTITUENT ELEMENTS → PRODUCTS.

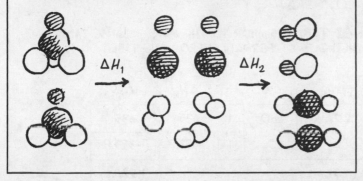

BREAKING THE REACTANTS INTO ELEMENTS HAS A HEAT CHANGE OF **MINUS** THE REACTANTS' TOTAL ENTHALPHY OF FORMATION:

$\Delta H_1 = $ −TOTAL ΔH_f OF ALL REACTANTS.

BUILDING THE PRODUCTS HAS A HEAT CHANGE EQUAL TO THE PRODUCTS' COMBINED ENTHALPHY OF FORMATION.

$\Delta H_2 = $ TOTAL ΔH_f OF ALL PRODUCTS.

THE ENTHALPY CHANGE OF THE ENTIRE REACTION, THEN, IS THE TOTAL ENTHALPY CHANGE OF THE TWO INTERMEDIATE REACTIONS:

$$\Delta H = \Delta H_1 + \Delta H_2$$
$$= \Delta H_f (PRODUCTS) - \Delta H_f (REACTANTS)$$

THAT IS, IN **ANY** REACTION, ΔH IS SIMPLY THE DIFFERENCE BETWEEN THE ENTHALPIES OF FORMATION OF THE PRODUCTS AND THE REACTANTS.

IT'S SO EASY!!

THIS, BY THE WAY, IS AN EXAMPLE OF A PRINCIPLE CALLED **HESS'S LAW:** ENTHALPY CHANGE DEPENDS ONLY ON THE BEGINNING AND END STATES, NOT ON ANYTHING IN BETWEEN. IF A REACTION HAS INTERMEDIATE STAGES, THEN ΔH IS THE SUM OF THE INTERMEDIATE ENTHALPY CHANGES.

IF NATURE SAYS TO SKIP THE IN-BETWEENS, WHO AM I TO DISOBEY NATURE?

HMM... YOU REALLY ARE LAZY!

Examples

LIMESTONE COOKS TO QUICKLIME:

$$CaCO_3(s) \xrightarrow{\Delta} CaO(s) + CO_2\uparrow \quad \Delta H = ?$$

WE MAKE AN **ENERGY-BALANCE TABLE,** SIMILAR TO THE MASS-BALANCE TABLES OF THE LAST CHAPTER. WE READ THE HEATS OF FORMATION FROM THE TABLE ON P. 100

REACTANT	n = no. of moles	ΔH_f	$n\Delta H_f$	PRODUCT	n	ΔH_f	$n\Delta H_f$
$CaCO_3$	1	-1207.6	-1207.6	CaO	1	-635	-635
				CO_2	1	-393.8	-393.8
TOTAL			-1,207.6				-1,028.8

THEN $\Delta H = \Delta H_f$ (PRODUCTS) $- \Delta H_f$ (REACTANTS)

$$= -1028.8 - (-1207.6) = 1207.6 - 1028.8$$

$$= 178.8 \text{ kJ FOR EACH MOLE OF } CaO \text{ PRODUCED.}$$

THE REACTION IS **ENDOTHERMIC,** AS WE HAVE SEEN.

TAKE THAT FROM THIS!

EXPLOSION OF NITROGLYCERINE:

$$4C_3H_5(NO_3)_3(l) \longrightarrow 6N_2\uparrow + O_2\uparrow + 12CO_2\uparrow + 10H_2O\uparrow$$

REACTANT	n	ΔH_f	$n\Delta H_f$	PRODUCT	n	ΔH_f	$n\Delta H_f$
$C_3H_5(NO_3)_3$	4	-364	-1456	N_2	6	0	0
				O_2	1	0	0
				H_2O (g)	10	-241.8	-2418.0
				CO_2 (g)	12	-393.8	-4725.6
TOTAL			-1456				-7143.6

$\Delta H = -7143.6 - (-1456) = -5687.6$ kJ FOR FOUR MOLES OF NITROGLYCERINE.

ONE MOLE OF NITRO RELEASES ONE-FOURTH AS MUCH:

$$\Delta H/\text{mole} = (-5687.6)/4 = -1421.9 \text{ kJ/mol.}$$

ONE MOLE OF NITROGLYCERINE WEIGHS 227 g, SO WE CAN ALSO CALCULATE ΔH/gram:

$$\Delta H/g = (-1421.9)/227 = -6.26 \text{ kJ/g.}$$

NOTE THAT NITRO-GLYCERINE RELEASED **TWICE** AS MUCH HEAT PER GRAM (6.26 kJ) AS BLACK POWDER (3.23 kJ).

COMBUSTION OF NATURAL GAS (METHANE, CH_4)

$$CH_4(g) + 2O_2(g) \longrightarrow CO_2(g) + 2H_2O(g)$$

REACTANT	n	ΔH_f	$n\Delta H_f$	PRODUCT	n	ΔH_f	$n\Delta H_f$
CH_4	1	−74.9	−74.9	$CO_2(g)$	1	−393.8	−393.8
				$H_2O(g)$	2	−241.8	−483.6
TOTAL			−74.9				−877.4

$\Delta H = -877.4 - (-74.9) = -802.5$ kJ/mol, OR ABOUT -50.2 kJ/g

WHEN O_2 IS THE OXIDANT IN A REDOX REACTION (AS ABOVE), THE ENTHALPY CHANGE IS CALLED THE **HEAT OF COMBUSTION.** COMBUSTION REACTIONS ARE HIGHLY EXOTHERMIC. BURNING HYDROGEN, FOR INSTANCE, RELEASES 286 kJ/mol OR 143 kJ/g. (= THE HEAT OF FORMATION OF WATER. SEE P. 100) SOME OTHER HEATS OF COMBUSTION, IN kJ PER GRAM OF FUEL:

HYDROGEN	143
NATURAL GAS (CH_4)	50
GASOLINE	48
CRUDE OIL	43
COAL	29
PAPER	20
DRIED BIOMASS	16
AIR-DRIED WOOD	15

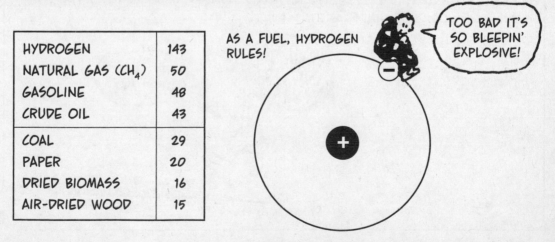

AS A FUEL, HYDROGEN RULES!

TOO BAD IT'S SO BLEEPIN' EXPLOSIVE!

IN THIS CHAPTER WE'VE SEEN HEAT CHANGES IN TWO DIFFERENT CONTEXTS: FIRST, ASSOCIATED WITH TEMPERATURE CHANGES, AND SECOND, ASSOCIATED WITH REACTIONS. IN THE NEXT CHAPTER, WE FIND HEAT IN ANOTHER, SURPRISING PLACE: CHANGES OF **STATE**.

YOU MEAN, LIKE GOING TO OREGON?

MOMO I

THAT IS, WHEN A SUBSTANCE CHANGES FROM A SOLID STATE TO LIQUID (OR LIQUID TO GAS, OR GAS TO SOLID, ETC.), HEAT IS ADDED OR TAKEN AWAY—AND THIS HAPPENS WITH NO CHANGE IN TEMPERATURE. AT TIMES, IN OTHER WORDS, HEAT CAN CHANGE **STRUCTURE** RATHER THAN TEMPERATURE.

HOW INEFFABLY MYSTERIOUS... WHERE DOES THE ENERGY GO?

TO UNDERSTAND THIS PUZZLE, WE NEED TO GO A BIT DEEPER INTO THE WORLD OF SOLIDS, LIQUIDS, AND GASES...

HEY, WAIT FOR ME!

Chapter 6
Matter in a State

UNDER ORDINARY CONDITIONS—OUTSIDE OF STARS, SAY—MATTER COMES IN THREE STATES: SOLID, LIQUID, AND GAS.

IN SOLIDS, PARTICLES ARE LOCKED TOGETHER IN A RIGID STRUCTURE. A SOLID HAS BOTH A DEFINITE SHAPE AND VOLUME.

IN LIQUIDS, PARTICLES CLING TOGETHER, BUT OVERALL STRUCTURE IS LACKING. A LIQUID HAS A DEFINITE VOLUME, BUT ITS SHAPE CONFORMS TO ITS CONTAINER.

IN GASES, STRUCTURE IS ABSENT. PARTICLES FLY AROUND ALMOST TOTALLY INDEPENDENTLY. A GAS HAS NEITHER A FIXED SHAPE NOR VOLUME, BUT WILL EXPAND TO FILL ANY CONTAINER.

AREN'T YOU GLAD SOLIDS AND LIQUIDS DON'T EXPAND TO FILL SPACE?

I DON'T KNOW... MIGHT BE FUN!

WHAT HOLDS SOLIDS AND LIQUIDS TOGETHER? THE ANSWER LIES WITH **INTERMOLECULAR FORCES** (IMFs) WITHIN THE SUBSTANCE. THESE ARE ATTRACTIONS BETWEEN MOLECULES (AS OPPOSED TO THE BONDS WITHIN A MOLECULE).

ONE IMF WE HAVE ALREADY ENCOUNTERED IS THE **HYDROGEN BOND.** IN WATER MOLECULES, ELECTRONS STAY CLOSER TO THE OXYGEN ATOM, SO THE HYDROGEN ATOMS EFFECTIVELY CARRY A POSITIVE CHARGE. THIS ATTRACTS THEM TO THE NEGATIVE POLE OF ANOTHER WATER MOLECULE.

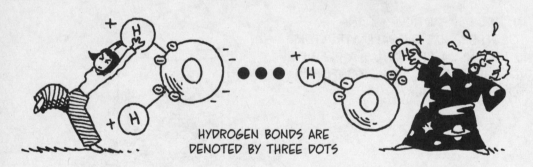

HYDROGEN BONDS ARE DENOTED BY THREE DOTS

BECAUSE OF ITS TWO ELECTRIC POLES, A WATER MOLECULE IS CALLED A **DIPOLE.** MANY OTHER MOLECULES ARE DIPOLES, TOO, AND THEY ATTRACT EACH OTHER END TO CHARGED END. DIPOLES MAY ALSO ATTRACT IONS.

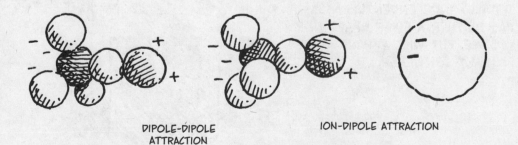

DIPOLE-DIPOLE ATTRACTION

ION-DIPOLE ATTRACTION

NONPOLAR MOLECULES CAN BECOME DIPOLES. FOR EXAMPLE, WHEN AN ION NEARS A MOLECULE, THE ION'S CHARGE CAN PUSH OR PULL THE MOLECULE'S ELECTRONS TOWARD ONE END. THE MOLECULE BECOMES AN **INDUCED DIPOLE,** AND ONE END IS ATTRACTED TO THE ION. A DIPOLE CAN INDUCE ANOTHER DIPOLE, TOO.

EVEN THE GHOSTLY FLIGHT OF ELECTRONS **WITHIN AN ATOM** OR MOLECULE CAN MAKE IT AN "INSTANTANEOUS" DIPOLE—WHICH CAN THEN INDUCE A NEARBY ATOM OR MOLECULE TO BECOME A DIPOLE, ETC. THE RESULTING RIPPLING ATTRACTION IS CALLED THE **LONDON DISPERSION FORCE.**

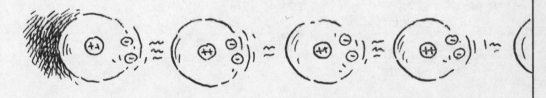

A TEMPORARY CHARGE IMBALANCE SETS OFF A RIPPLE OF DIPOLE-DIPOLE ATTRACTIONS.

ALTHOUGH THEY ARE CALLED INTER-**MOLECULAR** FORCES, THESE ATTRACTIONS DO NOT OPERATE ON MOLECULES ONLY. NOBLE GAS ATOMS, FOR INSTANCE, FEEL THE LONDON DISPERSION FORCE.

FROM NOW ON, WE'LL BE A LITTLE LOOSE WITH LANGUAGE AND SOMETIMES REFER TO IMFs AS BONDS. BONDS OR IMFs: THEY'RE ALL ELECTRIC ATTRACTIONS BE-TWEEN PARTICLES!

THIS TABLE SUMMARIZES THE STRENGTHS OF DIFFERENT ATTRACTIVE
FORCES. THE **STRENGTH** OF A BOND MEANS THE ENERGY REQUIRED
TO BREAK IT.

Strong attractions

	STRENGTH
IONIC ION-ION ATTRACTION	300 - 1000 kJ/mol
METALLIC ELECTRON SHARING AMONG METAL IONS	50 - 1000 kJ/mol
COVALENT ELECTRON SHARING	300 - 1000 kJ/mol

Moderate attractions

HYDROGEN BONDS AN EXPOSED PROTON IN ONE MOLECULE ATTRACTS A NEGATIVELY CHARGED ATOM IN A NEARBY MOLECULE	20 - 40 kJ/mol
ION-DIPOLE	10 - 20 kJ/mol

Weak attractions

DIPOLE-DIPOLE	1 - 5 kJ/mol
ION-INDUCED DIPOLE	1 - 3 kJ/mol
DIPOLE-INDUCED DIPOLE	0.05 - 2 kJ/mol
INSTANTANEOUS DIPOLE- INDUCED DIPOLE (DISPERSION)	0.05 - 2 kJ/mol

NOTE: DISPERSION FORCES
ARE GREATER BETWEEN
LARGER ATOMS, WHICH
HAVE MORE ELECTRONS TO
PUSH AROUND AND WHERE
ELECTRONS ARE FARTHER
FROM THE NUCLEUS AND
SO MORE EASILY PUSHED.

AS EVERYONE KNOWS WHO HAS EVER SEEN ICE MELT, **TEMPERATURE AFFECTS STATE.** RAISE THE TEMPERATURE OF ANYTHING HIGH ENOUGH, AND IT BECOMES A GAS. HOW HIGH DEPENDS ON THE BOND AND IMF STRENGTHS WITHIN THE SUBSTANCE.

WOW! A WATCHED POT REALLY DOES BOIL!

YOU'LL BE FAMOUS...

SUBSTANCES WITH WEAK IMFs CAN BE SOLID OR LIQUID ONLY AT VERY LOW TEMPERATURES, WHEN PARTICLES MOVE SLUGGISHLY.

AS TEMPERATURE RISES, MOLECULAR MOVEMENT STRAINS IMFs. IF THE FORCES ARE WEAK, THE SUBSTANCE MUST BECOME LIQUID OR GASEOUS.

SNAP!

BY CONTRAST, STRONGLY BONDED SUBSTANCES CAN REMAIN SOLID EVEN AT THOUSANDS OF DEGREES CELSIUS.

IN OTHER WORDS, SUBSTANCES WITH WEAK IMFs MELT AND BOIL AT LOWER TEMPERATURES, WHILE THOSE WITH STRONG BONDS MELT AND BOIL AT HIGHER TEMPERATURES. WATER, WITH ITS HYDROGEN BONDS, IS SOMEWHERE IN BETWEEN.

SUBSTANCE	FORCE	BOND STRENGTH (kJ/mol)	MELTING POINT (°C)	BOILING POINT (°C)
Ar	DISPERSION	8	−189	−186
NH_3	HYDROGEN	35	−78	−33
H_2O	HYDROGEN	23	0	100
Hg	METALLIC	68	−38	356
Al	METALLIC	324	660	2467
Fe	METALLIC	406	1535	2750
NaCl	IONIC	640	801	1413
MgO	IONIC	1000	2800	3600
Si	COVALENT	450	1420	2355
C (DIAMOND)	COVALENT	713	3550	4098

THE SIMPLEST STATE OF MATTER HAS (ALMOST) NO IMFS AT ALL.

Gases, Real and Ideal

GAS PARTICLES ZOOM AROUND FREELY, OR NEARLY SO. WHEN THEY DO BUMP INTO EACH OTHER, THEY FEEL AN IMF, SO THEIR COLLISIONS ARE A BIT "STICKY" (I.E., SOME K.E. IS LOST IN OVERCOMING THE ATTRACTION).

FOR THEORETICAL PURPOSES, CHEMISTS IGNORE THIS MINOR COMPLICATION AND THINK ABOUT AN **IDEAL GAS.** IN AN IDEAL GAS, ALL PARTICLES ARE IDENTICAL, THEY ZOOM AROUND FREELY, AND ALL COLLISIONS ARE PERFECTLY BOUNCY, OR **ELASTIC**—THAT IS, K.E. IS PRESERVED.

ONE CAN DISCUSS CERTAIN PROPERTIES OF AN IDEAL GAS:

n THE NUMBER OF MOLES, A MOLE BEING 6.02×10^{23} PARTICLES

V THE VOLUME

T THE TEMPERATURE IN DEGREES KELVIN

P THE PRESSURE

PRESSURE IS DEFINED AS **FORCE PER UNIT OF AREA.** A FORCE APPLIED TO A SMALL AREA CAN HAVE MORE EFFECT THAN A FORCE SPREAD OVER A LARGE AREA. THAT'S WHY YOU SIT ON A STOOL INSTEAD OF A NEEDLE! SAME FORCE (YOUR WEIGHT), DIFFERENT AREA.

GAS HAS PRESSURE BECAUSE ITS PARTICLES BUMP INTO THINGS.

SINCE DOUBLING AN AREA DOUBLES THE NUMBER OF COLLISIONS AND SO DOUBLES THE FORCE, FORCE AND AREA GO UP TOGETHER, SO THE PRESSURE IS CONSTANT THROUGHOUT THE GAS.

$$\text{Pressure} = \frac{\text{Force}}{\text{Area}}$$

THE AIR AROUND US EXERTS ATMOSPHERIC PRESSURE. **ONE ATMOSPHERE** (1 atm) IS THIS PRESSURE (ON AVERAGE) AT SEA LEVEL. IN TERMS OF METRIC UNITS:

$$1 \text{ atm} = 101{,}325 \text{ NEWTONS/m}^2$$
$$= 10.1325 \text{ NEWTONS/cm}^2$$

ATMOSPHERIC PRESSURE IS **HUGE!** WE DON'T FEEL IT BECAUSE IT PUSHES FROM ALL DIRECTIONS, BUT RECALL GUERICKE'S EXPERIMENT WITH HORSES TO APPRECIATE ITS TRUE MAGNITUDE.

MAYBE IF I HAD THE WIND AT MY BACK....

Gas Laws

NOT SURPRISINGLY, n, T, V, AND P ARE ALL RELATED. FOR INSTANCE, YOU MIGHT EXPECT THAT MORE PARTICLES WOULD OCCUPY A GREATER VOLUME, ALL ELSE BEING EQUAL. AND SO THEY DO! IN FACT, IT'S A LAW, THE FIRST OF THREE GAS LAWS, WHICH WE LIST IN ALPHABETICAL ORDER.

AVOGADRO'S LAW: IF T AND P ARE FIXED, THEN VOLUME IS PROPORTIONAL TO THE NUMBER OF MOLES.

$$\frac{n_1}{V_1} = \frac{n_2}{V_2}$$

OTHERWISE, PRESSURE WOULD CHANGE, WOULDN'T IT?

THIS IMPLIES THAT A SET VOLUME OF GAS (AT FIXED T AND P) ALWAYS HAS THE **SAME NUMBER OF MOLECULES**—NO MATTER WHAT WHAT GAS IT IS! THIS FACT ENABLED NINETEENTH-CENTURY CHEMISTS TO FIND ATOMIC WEIGHTS FOR THE FIRST TIME.

BOYLE'S LAW: IF n AND T ARE FIXED, THEN VOLUME IS INVERSELY PROPORTIONAL TO PRESSURE.

$$P_1 V_1 = P_2 V_2$$

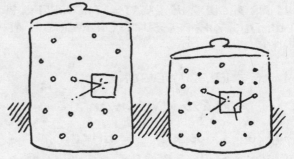

IN A LARGER VOLUME, FEWER PARTICLES HIT A UNIT OF AREA...

THAN IN A SMALLER VOLUME.

CHARLES'S LAW: WITH n AND P FIXED, VOLUME IS PROPORTIONAL TO TEMPERATURE.

$$\frac{V_1}{T_1} = \frac{V_2}{T_2}$$

IF T RISES...

MORE-ENERGETIC PARTICLES PUSH UP THE PISTON.

ALL THESE LAWS CAN BE ROLLED INTO A SINGLE EQUATION THAT COMBINES THE RELATIONSHIP AMONG ALL FOUR VARIABLES. IT'S CALLED THE **IDEAL GAS LAW,** AND IT GOES

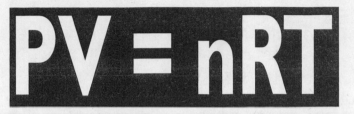

$$PV = nRT$$

R?

A CONSTANT OF NATURE.

HOLD ANY TWO VARIABLES FIXED, AND YOU SEE THE RELATIONSHIP BETWEEN THE OTHER TWO AS GIVEN IN THE A, B, C LAWS ON THE PREVIOUS PAGE.

R CAN BE FOUND AS FOLLOWS: FIRST, EXPERIMENTALLY DETERMINE THE VOLUME OF ONE MOLE OF GAS (ANY GAS, BY AVOGADRO!). AT $0°C$ (= $273°K$) AND 1 ATM, IT TURNS OUT THAT **ONE MOLE OF GAS OCCUPIES 22.4 LITERS.** SO:

n = 1 mol
T = 273°K
P = 1 atm
V = 22.4 L.

PLUG INTO THE GAS LAW EQUATION:

$$(1 \text{ atm})(22.4 \text{ L}) = (1 \text{ mol}) R (273°K)$$

SO

$$R = (22.4/273) \text{ atm-L/mol°K}$$
$$= 0.082 \text{ atm-L/mol°K}$$

THE CONDITIONS

T = 0°C AND
P = 1 atm

ARE KNOWN AS **STANDARD TEMPERATURE** AND **PRESSURE (STP).**

WHAT A GAS!

Example:

WHAT VOLUME OF GAS IS RELEASED BY THE EXPLOSION OF ONE GRAM OF BLACK POWDER?

$$4 KNO_3(s) + 7C(s) + S(s) \longrightarrow 3CO_2\uparrow + 3CO\uparrow + 2N_2\uparrow + K_2CO_3(s) + K_2S(s)$$

$$3 + 3 + 2 = 8 \text{ mol GAS}$$

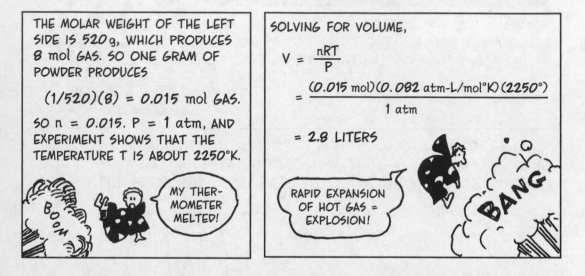

THE MOLAR WEIGHT OF THE LEFT SIDE IS 520g, WHICH PRODUCES 8 mol GAS. SO ONE GRAM OF POWDER PRODUCES

$$(1/520)(8) = 0.015 \text{ mol GAS.}$$

SO n = 0.015. P = 1 atm, AND EXPERIMENT SHOWS THAT THE TEMPERATURE T IS ABOUT 2250°K.

MY THER-MOMETER MELTED!

SOLVING FOR VOLUME,

$$V = \frac{nRT}{P}$$

$$= \frac{(0.015 \text{ mol})(0.082 \text{ atm-L/mol°K})(2250°)}{1 \text{ atm}}$$

$$= 2.8 \text{ LITERS}$$

RAPID EXPANSION OF HOT GAS = EXPLOSION!

BANG

A GRAM OF POWDER, WE MEASURE, OCCUPIES A TINY VOLUME, ABOUT 0.8 mL.

THE EVOLVED GAS EXPANDS TO (2800)/(0.8) = 3,500 TIMES THAT VOLUME! IF WE WANTED TO CONFINE THE GAS IN A LITTLE PACKAGE 1 mL (= .001 L) IN VOLUME, IT WOULD BUILD UP A PRESSURE OF:

$$P = \frac{nRT}{V}$$

$$= \frac{(0.015)(0.082)(2250)}{(0.001)}$$

OR ABOUT **2800** atm.

WHOA!

WHAT IS THIS UNHEALTHY OBSESSION WITH EXPLOSIONS?

Liquids

BECAUSE OF THEIR IMFs, LIQUIDS HAVE COMPLICATED BEHAVIOR. THERE ARE NO "IDEAL LIQUIDS."

BUT THERE ARE SOME VERY NICE ONES!

LIQUIDS BEHAVE AS IF THEY HAVE A SKIN. ATTRACTION AMONG SURFACE MOLECULES—**SURFACE TENSION**—KNITS THEM TOGETHER MORE TIGHTLY THAN INTERIOR MOLECULES. THAT EXPLAINS WHY BUGS CAN WALK ON WATER...

LIQUIDS EXPAND WHEN HEATED: AS MOLECULES MOVE FASTER, THEY GET FARTHER APART. THIS MAKES THERMOMETERS POSSIBLE: THE LIQUID—MERCURY OR WHATEVER—EXPANDS UP THE TUBE WHEN WARMED, AND SHRINKS WHEN COOLED.

AND WHY LIQUIDS FORM SPHERICAL DROPLETS!

Evaporation and Condensation

IN MOST LIQUIDS, MOLECULAR MOVEMENT CAN OVERCOME COHESIVE FORCES. IN THAT CASE, SOME MOLECULES BREAK THROUGH THE SURFACE AND **EVAPORATE.** CONVERSELY, LESS-ENERGETIC VAPOR MOLECULES MAY COLLECT INTO LIQUID, OR **CONDENSE.**

WHEN A MOLECULE GOES GASEOUS, ENERGY MUST BE ABSORBED FROM THE SURROUNDINGS TO BREAK THE ATTRACTIVE FORCES (BONDS, IMFs) THAT EXIST WITHIN THE LIQUID. **EVAPORATION IS ENDOTHERMIC.**

$$\text{liquid} \longrightarrow \text{gas} \quad \Delta H > 0$$

IN OTHER WORDS, GAS IS A **MORE ENERGETIC STATE OF MATTER** THAN LIQUID.

FOR EXAMPLE, WATER'S HEAT OF VAPORIZATION (AT 1 atm, 25°C) IS 44 kJ/mol. THAT IS THE ENTHALPY CHANGE OF THE "REACTION" $H_2O\,(l) \longrightarrow H_2O\,(g)$.

THIS IS WHY PERSPIRATION WORKS. EVAPORATING SWEAT DRAWS HEAT FROM YOUR BODY.

A BRILLIANTLY SIMPLE APPLICATION OF THIS 44 kJ/mol IS THE COOLING POT OF NIGERIAN POTTER **MOHAMMAD BAH ABBA.**

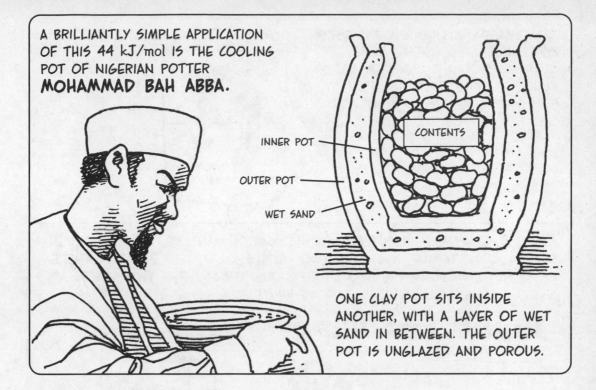

INNER POT

OUTER POT

WET SAND

CONTENTS

ONE CLAY POT SITS INSIDE ANOTHER, WITH A LAYER OF WET SAND IN BETWEEN. THE OUTER POT IS UNGLAZED AND POROUS.

WATER VAPOR AND HEAT

IN A DRY ENVIRONMENT, THE WATER IN THE SAND LAYER EVAPORATES AND PASSES OUT THROUGH PORES IN THE OUTER POT. IN THE PROCESS, IT DRAWS HEAT FROM THE APPARATUS.

THE TEMPERATURE INSIDE CAN FALL AS FAR AS 14°C (= 25°F) BELOW THAT OF THE OUTSIDE— A LIFESAVER IN DESERT COUNTRIES WHERE MOST PEOPLE CANNOT AFFORD A FRIDGE.

NOW IMAGINE A LIQUID IN A CLOSED CONTAINER AT CONSTANT TEMPERATURE. AS LIQUID EVAPORATES, VAPOR BUILDS UP, AND SOON SOME OF THIS VAPOR BEGINS TO CONDENSE.

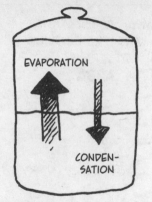

AT FIRST, EVAPORATION OUTPACES CONDENSATION, BUT EVENTUALLY, CONDENSATION MAY CATCH UP. WHEN THE TWO PROCESSES EXACTLY BALANCE, THERE IS NO NET CHANGE IN THE AMOUNT OF LIQUID OR GAS. THE TWO STATES ARE SAID TO BE IN **EQUILIBRIUM,** AND WE WRITE

liquid ⇌ vapor

NOTHING **APPEARS** TO BE HAPPENING, BUT ACTUALLY **TWO** THINGS ARE!

EQUAL RATES

THE EXTRA PRESSURE DUE TO VAPOR ALONE IS CALLED ITS **PARTIAL PRESSURE.*** AS VAPOR BUILDS UP, ITS PARTIAL PRESSURE RISES STEADILY (BIGGER n, SAME V AND T!) UNTIL EQUILIBRIUM. AT EQUILIBRIUM, THIS PARTIAL PRESSURE IS CALLED THE

vapor pressure.

IT'S THE PRESSURE THE VAPOR "WANTS" TO ATTAIN.

VAPOR PRESSURE (P_V) RISES WITH TEMPERATURE, SINCE MORE-AGITATED MOLECULES HAVE A GREATER "NEED" TO VAPORIZE.

WOW! TALK ABOUT DESIRE!

VAPOR PRESSURE OF WATER

T (°C)	P_V (ATM)
0	0.006
20	0.023
40	0.073
60	0.197
80	0.467
90	0.692
100	1.00
200	15.34
300	84.8

*THE TOTAL PRESSURE OF A MIXTURE OF GASES IS THE SUM OF ALL THEIR PARTIAL PRESSURES.

P_V IS THE PRESSURE AT WHICH VAPOR "WANTS" TO STABILIZE. BUT WHAT IF NO MATTER HOW MUCH VAPOR THE LIQUID SPEWS, ITS PRESSURE NEVER REACHES P_V? IN THAT CASE, VAPORIZATION GOES UNCHECKED, AND THE LIQUID **BOILS.**

WHAT ARE YOU DOING?

TRYING TO PUT THE VAPOR BACK IN THE POT!

WHETHER A LIQUID BOILS DEPENDS ON THE TOTAL PRESSURE ABOVE THE LIQUID—THE **EXTERNAL PRESSURE.** CALL IT P.

OW!

OW!

EQUILIBRIUM IS POSSIBLE WHEN VAPOR PRESSURE P_V **IS LESS THAN** P, BECAUSE THEN P_V CAN ACTUALLY BE REALIZED AS A PARTIAL PRESSURE OF VAPOR.

HERE H_2O MOLECULES ARE JUST PART OF THE AIR AND HAPPILY SO!

IF P IS LESS THAN P_V, THE PARTIAL PRESSURE OF VAPOR MUST ALSO BE LESS THAN P_V, AND BOILING OCCURS.

MORE!

MORE!

THAT IS, BOILING BEGINS PRECISELY WHEN **VAPOR PRESSURE EQUALS EXTERNAL PRESSURE.**

I JUST CAN'T PUSH DOWN HARD ENOUGH...

THE TEMPERATURE AT WHICH A LIQUID BOILS IS CALLED ITS

boiling point.

BOILING POINT DEPENDS ON EXTERNAL PRESSURE.

AT SEA LEVEL (PRESSURE = 1 ATM), WATER BOILS AT 100°C, BUT AT HIGH ALTITUDE, WHERE AIR IS THIN, BOILING POINT CAN DROP BELOW 85°. IN THE VACUUM OF SPACE, WATER BOILS AT ANY TEMPERATURE.

EXTERNAL PRESSURE, BY THE WAY, CAN INCLUDE LIQUID PRESSURE AS WELL AS GAS PRESSURE. IN THE DEEP OCEAN (PRESSURE VERY HIGH!) WATER NEAR VOLCANIC VENTS CAN REMAIN LIQUID ABOVE 350°C.

I'M COOKED!

WE SUMMARIZE ALL THIS WITH A LIQUID-GAS MINI-DIAGRAM. THE HORIZONTAL AXIS IS TEMPERATURE; THE VERTICAL AXIS IS PRESSURE; AND AT EACH PAIR OF VALUES (T,P) WE SEE WHETHER A SUBSTANCE IS LIQUID OR GAS.

THE CURVE BETWEEN THEM INDICATES THE BOILING POINT FOR ANY PRESSURE.

NOTE THAT PHASE TRANSITIONS CAN RESULT FROM CHANGING PRESSURE ALONE, OR TEMPERATURE ALONE, OR A COMBINATION.

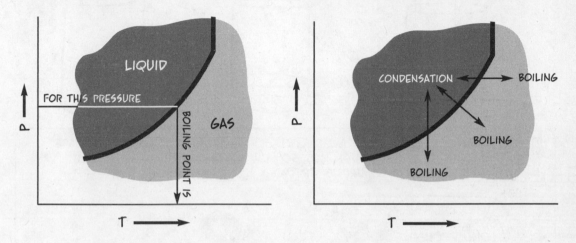

THE CURVE HAS ITS LIMITS. EVERY LIQUID HAS A CHARACTERISTIC **CRITICAL TEMPERATURE,** THE HIGHEST AT WHICH THE LIQUID STATE CAN EXIST. ABOVE THE CRITICAL TEMPERATURE, NO AMOUNT OF PRESSURE CAN STOP THE LIQUID FROM BOILING AWAY.

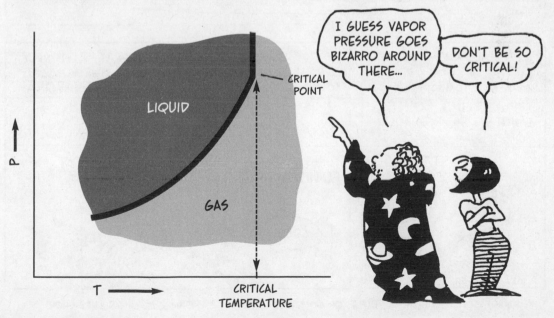

Melting Solids

IN THE OPEN AIR, MANY LIQUIDS SIMPLY EVAPORATE AWAY. SINCE THE VAPOR ESCAPES, IT BUILDS UP NO SIGNIFICANT PRESSURE ON THE SURFACE, AND EVAPORATION CONTINUES INDEFINITELY.

PARTIAL PRESSURE IS P_V AT SURFACE, BUT $<P_V$ HIGHER UP, SO...

MOLE-CULES KEEP LEAVING.

IN SOLIDS, BY CONTRAST, VERY FEW PARTICLES HAVE ENOUGH ENERGY TO ESCAPE. VAPOR PRESSURE IS LOW—THOUGH NOT SO LOW WE CAN'T SMELL MANY SOLIDS. IN SOME CASES, VAPOR PRESSURE IS VIRTUALLY NIL. DIAMONDS ARE FOREVER!

WHO KNOWS? MAYBE IF WE WAIT LONG ENOUGH...

AS WE ALL KNOW, SOLIDS **MELT***, AND THEY DO SO AT A SET TEMPERATURE, THE **MELTING POINT**, WHICH VARIES FROM SOLID TO SOLID.

WHAT'S THE MELTING POINT OF ROAST BEEF?

AT THIS TEMPERATURE, ANY ADDED HEAT IS ENTIRELY CONSUMED IN BREAKING BONDS UNTIL THE SOLID IS COMPLETELY MELTED. **MELTING, LIKE EVAPORA-TION, IS ENDOTHERMIC.**

$$\text{SOLID} \longrightarrow \text{LIQUID} \quad \Delta H > 0$$

THIS ENTHALPHY CHANGE IS CALLED THE **HEAT OF FUSION.** FOR ICE AT STP, IT'S 6.01 kJ/mol.

COOL!

*USUALLY. SOME OF THEM **SUBLIME,** OR GO STRAIGHT TO THE GAS PHASE. MORE ON THAT SHORTLY.

EXTERNAL PRESSURE AFFECTS MELTING POINT: IN THIS SOLID-LIQUID MINI-DIAGRAM WITH P AND T AXES, THE CURVE SHOWS THE MELTING POINT FOR EACH VALUE OF P.

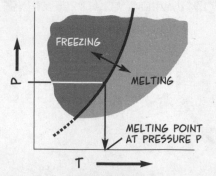

FREEZING

MELTING

P

MELTING POINT AT PRESSURE P

T

THE EFFECT IS LESS DRAMATIC THAN WITH BOILING POINT, HOWEVER, SO THE MELTING CURVE IS USUALLY PRETTY STEEP.

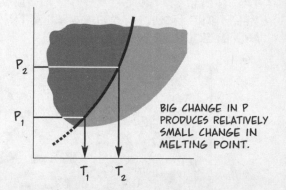

P_2

P_1

BIG CHANGE IN P PRODUCES RELATIVELY SMALL CHANGE IN MELTING POINT.

T_1 T_2

IN A FEW WEIRD MATERIALS, ADDED PRESSURE ACTUALLY DECREASES MELTING POINT. WATER IS ONE SUCH.

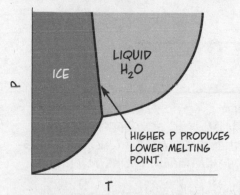

ICE

LIQUID H_2O

P

T

HIGHER P PRODUCES LOWER MELTING POINT.

THAT'S BECAUSE WATER **EXPANDS** WHEN IT FREEZES. THE CRYSTALLINE STRUCTURE OF ICE IS UNUSUALLY SPACIOUS.

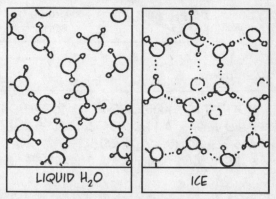

LIQUID H_2O ICE

PRESSING ON AN ICE CUBE PUTS STRAIN ON THE BONDS AND DRIVES THE MOLECULES INTO A TIGHTER BUT MORE RANDOM CONFIGURATION, AND THE ICE MELTS AT THE POINT OF PRESSURE.

SO, UNLIKE MOST SOLIDS, ICE FLOATS ON ITS LIQUID FORM... THE EXPANSION OF FREEZING WATER CAN CRACK ROCKS... AND THIS ODD FEATURE HAS A PROFOUND IMPACT ON THE WORLD AROUND US.

ICE-SKATING AS IT WOULD BE IF WATER FROZE LIKE A NORMAL SUBSTANCE.

Phase Diagrams

PUT OUR MINI-DIAGRAMS TOGETHER AND THEY SHOW A COMPLETE PICTURE OF
THE THREE STATES OF MATTER IN TERMS OF T AND P. THE SOLID-LIQUID CURVE
MEETS THE LIQUID-GAS CURVE AT A **TRIPLE POINT** WHERE ALL THREE PHASES
ARE IN EQUILIBRIUM.

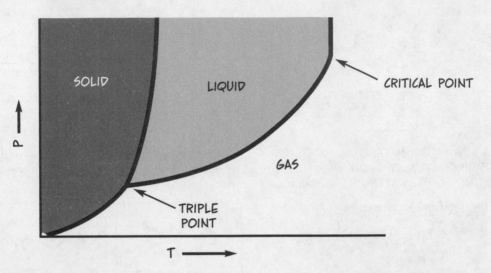

NOTE THAT THERE ARE ALSO CONDITIONS WHEN A SOLID CAN CHANGE DIRECTLY
INTO A GAS, A PROCESS CALLED **SUBLIMATION**. THE REVERSE PROCESS,
GAS → SOLID, IS **DEPOSITION**. THE BEST-KNOWN EXAMPLE AT NORMAL PRES-
SURE IS CO_2, "DRY ICE," THE STUFF USED IN THEATRICAL SMOKE MACHINES.

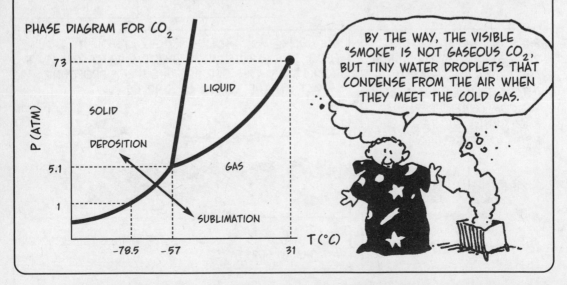

PHASE DIAGRAM FOR CO_2

BY THE WAY, THE VISIBLE
"SMOKE" IS NOT GASEOUS CO_2,
BUT TINY WATER DROPLETS THAT
CONDENSE FROM THE AIR WHEN
THEY MEET THE COLD GAS.

A COUPLE OF OTHER PHASE DIAGRAMS SHOW SOME MORE SUBTLE AND UNUSUAL FEATURES OF MATTER. HERE IS **CARBON.**

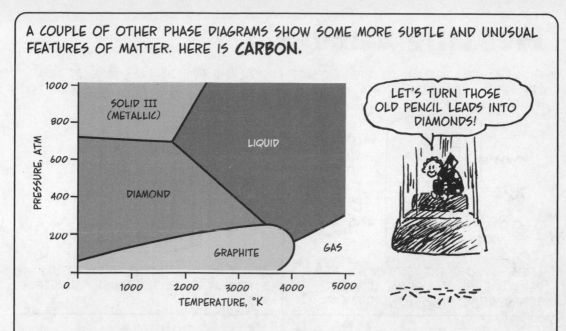

CARBON HAS **THREE SOLID FORMS,** WITH DIFFERENT CRYSTALLINE STRUCTURES: GRAPHITE, FOUND IN COAL AND PENCIL LEADS, DIAMOND, WHICH IS FORMED ONLY UNDER HIGH-PRESSURE CONDITIONS, AND METALLIC, WHICH EXISTS ONLY AT EXTREMELY HIGH PRESSURE. NOTE HOW THE MELTING CURVE SLOPES DIFFERENTLY FOR EACH TYPE OF CRYSTAL.

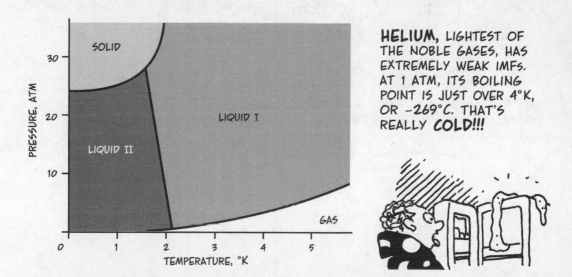

HELIUM, LIGHTEST OF THE NOBLE GASES, HAS EXTREMELY WEAK IMFs. AT 1 ATM, ITS BOILING POINT IS JUST OVER 4°K, OR −269°C. THAT'S REALLY **COLD!!!**

BELOW THAT TEMPERATURE IT IS A LIQUID... AND BELOW 2.17°K—IT IS ANOTHER KIND OF LIQUID! THIS HELIUM II IS A "SUPERFLUID" WITH WEIRD PROPERTIES. IT FLOWS WITHOUT VISCOSITY (GOOPINESS)... IT WILL LEAK OUT THE TINIEST PORE... IT WILL EVEN CLIMB THE CONTAINER WALLS! SEE http://cryowwwebber.gsfc.nasa.gov/introduction/liquid_helium.html FOR DETAILS. HELIUM CAN ALSO BE SOLID, BUT ONLY AT PRESSURES ABOVE 25 ATM.

Heating Curves

FINALLY, LET'S RETURN TO THE HEATS OF FUSION AND EVAPORATION, AND SEE HOW THEY PLAY OUT WHEN WE HEAT A BLOCK OF ICE UNTIL IT MELTS AND THEN BOILS.

LET'S USE MICROWAVES TO HEAT THE WATER UNIFORMLY.

LET'S SUPPOSE THE ICE'S INITIAL TEMPERATURE IS –5°C. AS WE ADD HEAT, TEMPERATURE RISES TOWARD 0°C.

AT THE MELTING POINT, TEMPERATURE STALLS AT 0°, EVEN THOUGH WE KEEP ADDING HEAT.

ALL THE ADDED HEAT GOES INTO BREAKING BONDS WITHIN THE ICE CRYSTAL.

ONLY WHEN THE ICE IS FULLY MELTED DOES TEMPERATURE RISE AGAIN.

AT THE BOILING POINT, TEMPERATURE AGAIN STALLS, AS HEAT IS TAKEN UP BY PHASE CHANGE ALONE.

ONCE THE WATER IS FULLY VAPORIZED, THE STEAM'S TEMPERATURE RISES.

THAT SIX-PANEL COMIC STRIP TRANSLATES INTO THIS **HEATING CURVE** THAT PLOTS TEMPERATURE AGAINST ADDED HEAT. T STOPS RISING DURING PHASE TRANSITIONS.

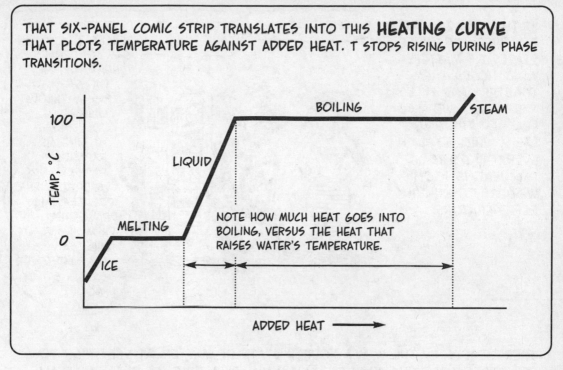

NOTE HOW MUCH HEAT GOES INTO BOILING, VERSUS THE HEAT THAT RAISES WATER'S TEMPERATURE.

THE SPECIFIC HEAT OF WATER, RECALL, IS AROUND 4.18 J/g°C . SO TO RAISE THE TEMPERATURE OF ONE GRAM OF LIQUID WATER BY 100° REQUIRES AN ADDITION OF ABOUT

$$(4.18 \text{ J}/\,°C)(100°C)$$

$$= 418 \text{ Joules}$$

BY CONTRAST, AT 100°C THE HEAT OF VAPORIZATION OF WATER IS ABOUT 41 **KILO**JOULES PER MOLE. SINCE A MOLE OF WATER WEIGHS 18 GRAMS, THIS IS

$$\frac{41 \text{ kJ/mol}}{18 \text{ g/mol}} = 2.28 \text{ kJ/g}$$

$$= 2{,}280 \text{ Joules/gram}$$

THAR'S JOULES IN THEM THAR HYDROGEN BONDS!

SHOOT, AN' I NEVER TOUCH TH' STUFF.

IN OTHER WORDS, IT TAKES ABOUT **FIVE TIMES AS MUCH HEAT** TO BOIL WATER COMPLETELY AWAY AS IT DOES TO HEAT IT ALL THE WAY FROM 0° TO 100°!!

IN THIS CHAPTER, WE REVIEWED THE THREE STATES OF MATTER, WHAT HOLDS THEM TOGETHER AND PULLS THEM APART. WE ALSO LEARNED THE GAS LAWS, WHICH EXPLAIN EVERYTHING FROM CALCULATING ATOMIC WEIGHTS TO RUNNING REFRIGERATORS.

REFRIGERATORS?

YES... ELECTRICITY DRIVES A PUMP... THE PUMP COMPRESSES A GAS... THE GAS CONDENSES... HEATS UP, BY THE GAS LAWS... PASSES THROUGH COILS... IS COOLED BY OUTSIDE AIR... EXPANDS RAPIDLY AND VAPORIZES... ENDOTHERMICALLY DRAWS HEAT FROM INSIDE THE... SAY... IS THAT LEFTOVER SALAMI STILL GOOD?

THERE EXISTS, BY THE WAY, A FOURTH STATE OF MATTER. AT VERY HIGH TEMPERATURE, ELECTRONS JUMP OFF THEIR NUCLEI; ALL BONDS BREAK; AND ALL SUBSTANCES TURN INTO A HOT PARTICLE SOUP CALLED **PLASMA.** LUCKILY, THIS IS NOT SOMETHING CHEMISTS HAVE TO THINK ABOUT VERY OFTEN...

WE ONLY LIKE STATES OF MATTER WE CAN EAT, DRINK, OR BREATHE!

Chapter 7
Solutions

WE'VE JUST LOOKED AT STATES OF MATTER ONE AT A TIME... NOW LET'S COMBINE TWO OF THEM— OR RATHER, LET'S COMBINE SOMETHING, ANYTHING, WITH A LIQUID. FOR INSTANCE: ADD A PINCH OF TABLE SALT TO A FLASK OF WATER.

THE SALT, OF COURSE, COMPLETELY VANISHES.

THE SALT, AS WE SAY, **DISSOLVES** IN THE WATER.

SAY, WHERE'D YOU COME FROM, ANYWAY?

THE MAGIC OF CARTOONING.

WHEN A SUBSTANCE DISSOLVES IN A LIQUID, THE COMBINATION IS CALLED A **SOLUTION**. THE LIQUID IS THE **SOLVENT**, AND THE DISSOLVED MATERIAL IS THE **SOLUTE**.*

Solute + Solvent → Solution

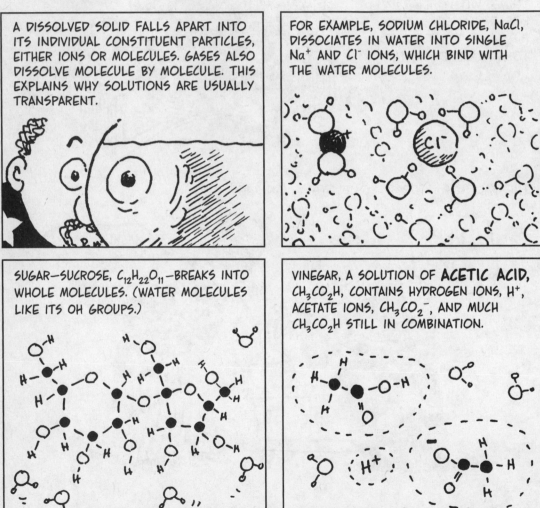

A DISSOLVED SOLID FALLS APART INTO ITS INDIVIDUAL CONSTITUENT PARTICLES, EITHER IONS OR MOLECULES. GASES ALSO DISSOLVE MOLECULE BY MOLECULE. THIS EXPLAINS WHY SOLUTIONS ARE USUALLY TRANSPARENT.

FOR EXAMPLE, SODIUM CHLORIDE, $NaCl$, DISSOCIATES IN WATER INTO SINGLE Na^+ AND Cl^- IONS, WHICH BIND WITH THE WATER MOLECULES.

SUGAR—SUCROSE, $C_{12}H_{22}O_{11}$—BREAKS INTO WHOLE MOLECULES. (WATER MOLECULES LIKE ITS OH GROUPS.)

VINEGAR, A SOLUTION OF **ACETIC ACID**, CH_3CO_2H, CONTAINS HYDROGEN IONS, H^+, ACETATE IONS, $CH_3CO_2^-$, AND MUCH CH_3CO_2H STILL IN COMBINATION.

*ACTUALLY, A SOLUTION CAN BE SOLID OR GASEOUS TOO. ANY HOMOGENEOUS MIXTURE OF TWO OR MORE SUBSTANCES IS CONSIDERED A SOLUTION, WHATEVER ITS PHASE.

LET'S LOOK MORE CLOSELY AT THE DISSOLVING PROCESS. IMAGINE A CHUNK OF MATERIAL IMMERSED IN LIQUID. IN ORDER TO DISSOLVE, SOME OF ITS PARTICLES MUST BREAK THE BONDS THAT HOLD THEM TOGETHER AND FORM NEW BONDS WITH MOLECULES OF LIQUID. SIMILARLY, IMFs WITHIN THE LIQUID MUST ALSO BE OVERCOME.

EACH FREE SOLUTE PARTICLE ATTRACTS ONE OR MORE MOLECULES OF SOLVENT, WHICH CLUSTER AROUND IT IN A SOLVENT "CAGE." THIS PROCESS OF BREAKING AND FORMING BONDS IS CALLED **SOLVATION.**

ALL THIS BOND REARRANGING MEANS THAT **DISSOLVING IS A CHEMICAL REACTION.** AMONG OTHER THINGS, THEN, IT HAS AN ASSOCIATED ENTHALPY CHANGE, WHICH MAY BE POSITIVE OR NEGATIVE.

ΔH AGAIN!

FOR EXAMPLE, WHEN MAGNESIUM CHLORIDE, $MgCl_2$, DISSOLVES IN WATER, IT HAS AN ENTHALPY OF SOLVATION

$$\Delta H = 119 \ kJ/mol$$

HIGHLY ENDOTHERMIC! A MERE 4g OF $MgCl_2$ (= .042 mol) IN 50 mL (= 50 g) OF WATER DROPS THE WATER'S TEMPERATURE BY 23.9°C (BY THE BASIC CALORIMETRY EQUATION).

CHEMICAL COLD PACKS ARE IN FACT MADE FROM $MgCl_2$ AND OTHER SALTS THAT ABSORB HEAT WHEN DISSOLVED IN WATER.

MORE ON ENERGY IN THE NEXT CHAPTER... BUT FOR NOW... I'M CHILLIN'... AHHH...

SOME LIQUID MIXTURES ARE NOT SOLUTIONS:

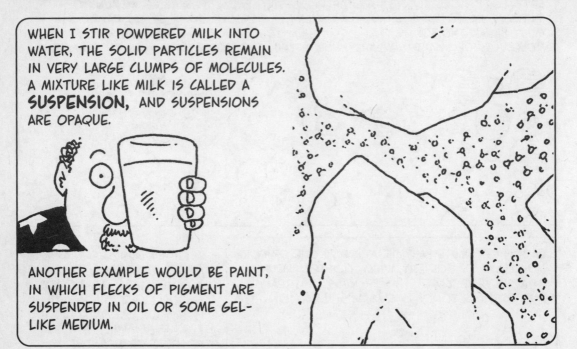

WHEN I STIR POWDERED MILK INTO WATER, THE SOLID PARTICLES REMAIN IN VERY LARGE CLUMPS OF MOLECULES. A MIXTURE LIKE MILK IS CALLED A **SUSPENSION,** AND SUSPENSIONS ARE OPAQUE.

ANOTHER EXAMPLE WOULD BE PAINT, IN WHICH FLECKS OF PIGMENT ARE SUSPENDED IN OIL OR SOME GEL-LIKE MEDIUM.

AN **EMULSION** IS A SUSPENSION OF ONE LIQUID IN ANOTHER. MAYONNAISE, FOR EXAMPLE, MAINLY CONSISTS OF TINY DROPLETS OF OIL SUSPENDED IN VINEGAR. ORDINARILY, OIL AND VINEGAR WOULD SEPARATE, BUT THE ADDITION OF A SMALL AMOUNT OF MUSTARD AND EGG YOLK STABILIZES THE EMULSION.

LONG MOLECULES FROM THE YOLK BURROW INTO OIL DROPLETS. A POLAR "TAIL" STICKS OUT AND ATTRACTS THE POLAR WATER MOLECULES IN VINEGAR, WHICH BLOCK THE DROPLETS FROM MERGING.

FROM NOW ON, WE CONCENTRATE ON SOLUTIONS.

CONCENTRATE— GET IT?

I'M AFRAID SO...

Concentration

IS A MEASURE OF HOW MUCH SOLUTE IS PRESENT IN A SOLUTION RELATIVE TO THE WHOLE.

FOR EXAMPLE, WEIGH OUT 35 g OF NaCl. PUT IT IN A GRADUATED CONTAINER AND ADD WATER UNTIL THERE IS ONE LITER OF SOLUTION.

THE CONCENTRATION OF THIS SOLUTION IS 35 g/L AND MEASURES **MASS OF SOLUTE** PER **VOLUME OF SOLUTION.**

OTHER POSSIBLE MEASURES (ALL USED!):

  MASS OF SOLUTE PER MASS OF SOLUTION

  VOLUME OF SOLUTE PER VOLUME OF SOLUTION

  MASS OF SOLUTE PER VOLUME OF SOLVENT
    (NOT THE SAME THING AS VOLUME OF SOLUTION!)

  MASS OF SOLUTE PER MASS OF SOLVENT

  PARTS PER MILLION (PPM)
    (A MASS-PER-MASS RATIO OF VERY DILUTE SOLUTIONS)

  PARTS PER BILLION (PPB, EVEN MORE DILUTE)

IT'S GOOD TO HAVE OPTIONS!

WHEN THE SOLVENT IS WATER, WE CAN EASILY CONVERT FROM A MASS-VOLUME RATIO TO A MASS-MASS RATIO, BECAUSE **ONE LITER OF WATER WEIGHS ONE KILOGRAM.** A LITER OF VERY DILUTE AQUEOUS SOLUTION, OF COURSE, WEIGHS THE SAME.

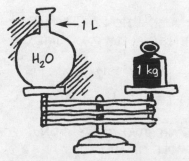

OUR FAVORITE MEASURE OF CONCENTRATION ACTUALLY TELLS YOU HOW MANY MOLECULES ARE DISSOLVED RELATIVE TO VOLUME. **MOLARITY,** OR MOLAR CONCENTRATION, IS THE NUMBER OF MOLES OF SOLUTE PER LITER OF SOLUTION. WE WRITE

M = MOLES/LITER.

RATS!

NO, MOLES!

AH! IT'S A MEASURE OF **GOODNESS** THEN!

SORRY, IT'S **MOLARITY,** NOT MORALITY...

WHAT'S THE MOLARITY OF OUR 35 g/L SALT SOLUTION? ONE MOLE OF NaCl WEIGHS 58.4 g, SO WE HAVE

$$\frac{35 \text{ g}}{58.4 \text{ g/mol}} = 0.6 \text{ mol NaCl}$$

IN A LITER OF SOLUTION. THE MOLARITY IS 0.6 M.

WE USE SQUARE BRACKETS, [ ], TO DENOTE MOLAR CONCENTRATION OF ANY "SPECIES" (I.E., ANY PARTICULAR MOLECULE OR ION) IN SOLUTION. HERE, SINCE NaCl DISSOCIATES COMPLETELY IN SOLUTION,

$$[Na^+] = 0.6M$$
$$[Cl^-] = 0.6M$$

IN A 1 M SOLUTION OF Na_2SO_4, WHICH ALSO FULLY DISSOCIATES,

$$[Na^+] = 2 \text{ M}$$
$$[SO_4{}^{2-}] = 1 \text{ M}$$

THERE ARE TWO MOLES OF Na^+ FOR EACH MOLE OF Na_2SO_4.

HMM... I WONDER HOW MUCH A MOLE OF MOLES WEIGHS...

Solubility

ANY SUBSTANCE WILL DISSOLVE IN ANY LIQUID—TO SOME DEGREE, THOUGH IT MAY BE VERY SMALL INDEED. FOR INSTANCE, NO MORE THAN .000006 g OF ELEMENTAL MERCURY (Hg) WILL DISSOLVE IN A LITER OF WATER AT ROOM TEMPERATURE. A MOLE OF Hg WEIGHS 200.6 g...

SO [Hg] IS ONLY... UM...

$$\frac{.000006}{200.6}$$

3×10^{-8} M !!

BUT EVEN WHEN A SUBSTANCE IS HIGHLY SOLUBLE, THERE IS ALWAYS A LIMIT! YOU CAN THROW ONLY SO MUCH SALT INTO WATER BEFORE IT STARTS PILING UP ON THE BOTTOM, UNDISSOLVED.

AND NOW IT **REALLY** TASTES **HORRIBLE.**

THIS LIMIT, A SUBSTANCE'S MAXIMUM POSSIBLE CONCENTRATION, IS CALLED ITS **SOLUBILITY.** A MAXIMALLY CONCENTRATED SOLUTION IS CALLED **SATURATED.**

WE SAY A MATERIAL IS SOLUBLE IF IT DISSOLVES TO AN "APPRECIABLE" DEGREE, AND INSOLUBLE IF NOT—A FUZZY CONCEPT, CLEARLY.

THE EQUIVALENT WORD FOR LIQUID-LIQUID INTERACTION IS **MISCIBILITY:** TWO LIQUIDS ARE MISCIBLE IF THEY DISSOLVE IN ONE ANOTHER AND IMMISCIBLE IF, LIKE OIL AND WATER, THEY SEPARATE.

OIL

WATER

FOOD COLORING

WATER

IMMISCIBLE

MISCIBLE

SOME FACTORS AFFECTING SOLUBILITY:

LIKE TENDS TO DISSOLVE LIKE. A POLAR SOLVENT (SUCH AS WATER) TENDS TO DISSOLVE (OR MIX WITH) OTHER POLAR COMPOUNDS. HERE DIPOLE-DIPOLE OR DIPOLE-ION ATTRACTIONS DRIVE SOLVATION. FOR INSTANCE:

METHANOL, CH_3OH, IS POLAR AND FORMS A HYDROGEN BOND WITH WATER, WITH WHICH IT WILL MIX IN ANY AMOUNT.

ITS COUSIN **METHANE,** CH_4, IS UTTERLY SYMMETRICAL AND NONPOLAR. WATER SHUNS IT, AND ITS SOLUBILITY IS VERY LOW (0.024 g/L OR 0.0015 M).

I JUST DON'T FIND IT ATTRACTIVE IN THE LEAST...

MOLECULAR SIZE:
BIG, HEAVY MOLECULES TEND TO BE LESS SOLUBLE THAN SMALL, LIGHT ONES. SOLVENT MOLECULES FIND IT HARD TO "CAGE" BIG PARTICLES.

I MEAN, WHERE DO YOU START?

TEMPERATURE

ALSO AFFECTS SOLUBILITY. AS TEMPERATURE RISES, AGITATED MOLECULES OR IONS BREAK THEIR BONDS MORE EASILY, SO SOLUBILITY USUALLY GOES UP. EXCEPTIONS EXIST, HOWEVER, AND THE EFFECT IS SOME- TIMES SLIGHT.

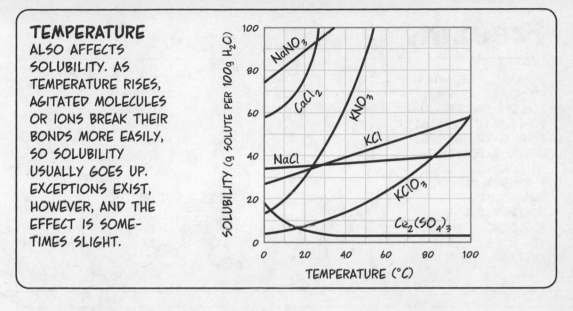

FOR DISSOLVED GASES, PRESSURE AFFECTS SOLUBIL- ITY. TO BE PRECISE, THE **PARTIAL PRESSURE** OF A GAS ABOVE THE SOLUTION AFFECTS THE AMOUNT OF GAS THAT WILL DIS- SOLVE. THE HIGHER THE PARTIAL PRESSURE, THE GREATER THE GAS'S SOLUBILITY.

LOWER PRESSURE
LOWER CONCENTRATION

HIGHER PRESSURE
HIGHER CONCENTRATION

AND THIS IS SUPPOSED TO BE A GOOD THING...

SOFT DRINKS, WHICH CONTAIN DISSOLVED CO_2, ARE BOTTLED AT HIGH PRESSURE TO INCREASE THE AMOUNT OF DISSOLVED GAS. WHEN THE CAP IS REMOVED, PRESSURE EASES, AND CO_2 FIZZES OUT OF SOLUTION.

Freezing

GENERALLY SPEAKING, DISSOLVED MATERIAL LOWERS THE FREEZING POINT. SOLUTE PARTICLES DISRUPT THE NORMAL COHESIVE FORCES WITHIN THE SOLVENT, MAKING IT HARDER FOR THE SOLUTION TO SOLIDIFY. THE HIGHER THE CONCENTRATION, THE LOWER THE FREEZING POINT.

EEK! I'M BEING PULLED DOWN BY MOLES!

IT'S **SO** HARD TO CRYSTALLIZE SOMETIMES...

FOR EXAMPLE, IN AN ICE CREAM MAKER, A BUCKET OF CREAM, DISSOLVED SUGAR, AND FLAVOR IS SURROUNDED BY ICE, WHICH MAY BE AT −3° TO −5°C.

WHEN SALT IS ADDED, THE ICE MELTS. THE BELOW-ZERO SALT WATER NOW MAKES CONTACT WITH THE FULL SURFACE OF THE BUCKET.

NOW THE CREAM CAN BE RAPIDLY COOLED BELOW 0°C. LIQUID WATER ALSO HAS A HIGHER HEAT CAPACITY THAN ICE, AND SO COOLS MORE EFFICIENTLY.

LIQUID CREAM

ICE TOUCHES THE CREAM CONTAINER IN ONLY A FEW PLACES.

ICE CREAM

EFFICIENT HEAT TRANSFER

ICE CREAM RARELY FREEZES TOTALLY. AS THE LIQUID FREEZES, SUGAR BECOMES MORE CONCENTRATED IN THE REMAINING SYRUP, SO ITS FREEZING POINT DROPS EVEN LOWER, AND SOME OF IT REMAINS UNFROZEN. THAT'S WHY ICE CREAM IS USUALLY SOFT.

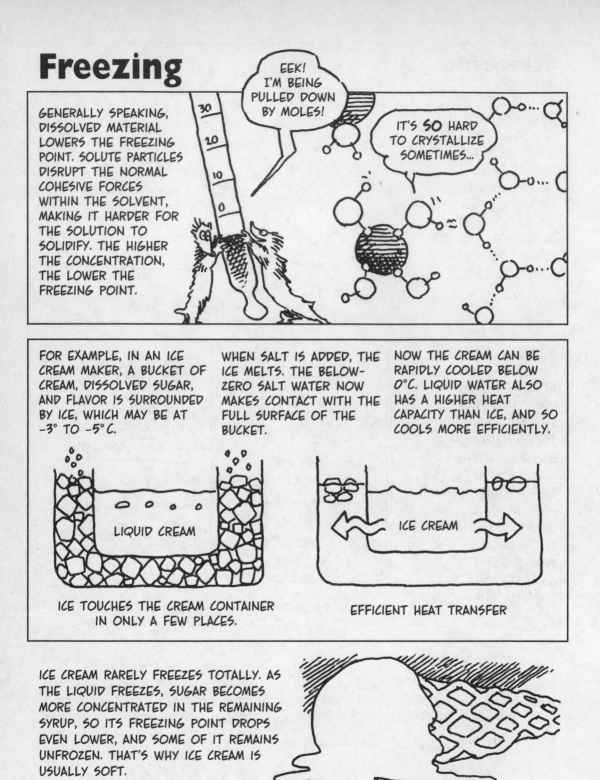

138

Boiling

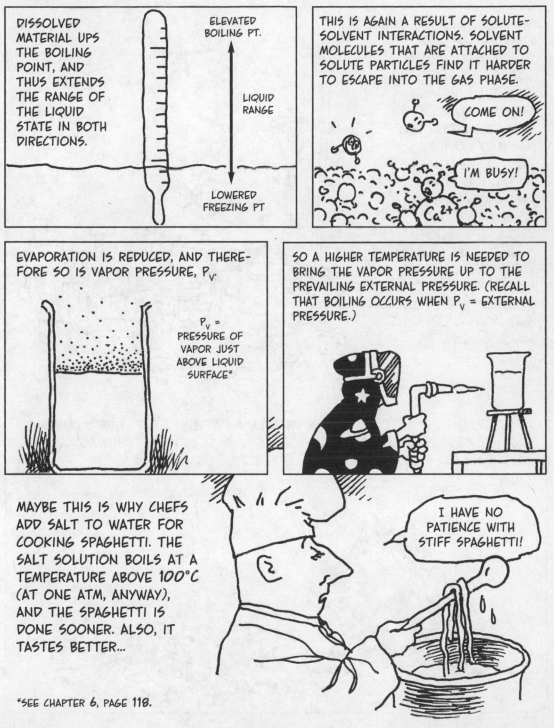

DISSOLVED MATERIAL UPS THE BOILING POINT, AND THUS EXTENDS THE RANGE OF THE LIQUID STATE IN BOTH DIRECTIONS.

ELEVATED BOILING PT.

LIQUID RANGE

LOWERED FREEZING PT

THIS IS AGAIN A RESULT OF SOLUTE-SOLVENT INTERACTIONS. SOLVENT MOLECULES THAT ARE ATTACHED TO SOLUTE PARTICLES FIND IT HARDER TO ESCAPE INTO THE GAS PHASE.

COME ON!

I'M BUSY!

Ca^{2+}

EVAPORATION IS REDUCED, AND THERE-FORE SO IS VAPOR PRESSURE, P_V.

P_V = PRESSURE OF VAPOR JUST ABOVE LIQUID SURFACE*

SO A HIGHER TEMPERATURE IS NEEDED TO BRING THE VAPOR PRESSURE UP TO THE PREVAILING EXTERNAL PRESSURE. (RECALL THAT BOILING OCCURS WHEN P_V = EXTERNAL PRESSURE.)

MAYBE THIS IS WHY CHEFS ADD SALT TO WATER FOR COOKING SPAGHETTI. THE SALT SOLUTION BOILS AT A TEMPERATURE ABOVE 100°C (AT ONE ATM, ANYWAY), AND THE SPAGHETTI IS DONE SOONER. ALSO, IT TASTES BETTER...

I HAVE NO PATIENCE WITH STIFF SPAGHETTI!

*SEE CHAPTER 6, PAGE 118.

So What?

AN ENORMOUS AMOUNT OF FAMILIAR AND IMPORTANT CHEMISTRY HAPPENS IN SOLUTION: COOKING, BREWING, FERMENTATION, DIGESTION, ELECTRIC BATTERY POWER, MEDICINE, ETCHING OF METAL AND GLASS, LAUNDRY AND OTHER WASHING, BLOOD CHEMISTRY, TOOTH DECAY, CALCIFICATION OF PIPES, ACID RAIN, OIL REFINING, WATER PURIFICATION, CELLULAR METABOLISM—JUST TO NAME A FEW!

MUCH OF THE REST OF THIS BOOK WILL BE AN ATTEMPT TO UNDERSTAND SUCH PROCESSES IN MORE DETAIL. WE BEGIN BY LOOKING AT WHY SOME REACTIONS GO FAST, WHILE OTHERS GO SLOW...

Chapter 8
Reaction Rate and Equilibrium

IN CHEMISTRY, WE CARE ABOUT NOT ONLY **WHAT** REACTS, BUT ALSO **HOW FAST.** BLACK POWDER EXPLODES IN A FLASH, WHILE THE SUGAR IN YOUR COFFEE NEVER SEEMS TO DISSOLVE FAST ENOUGH. WE TRY TO SPEED UP ENVIRONMENTAL CLEANUP AND RETARD RUST AND AGING. IN OTHER WORDS, **RATES MATTER!**

I'D LIKE TO SPEED UP **HIS** RATE!

"AT FIRST SIGHT, NOTHING SEEMS MORE OBVIOUS THAN THAT EVERYTHING HAS A BEGINNING AND AN END."

—SVANTE ARRHENIUS, 1903 NOBEL PRIZE WINNER IN CHEMISTRY

WHAT'S THE RATE OF A CHEMICAL REACTION? WE BEGIN WITH THE ULTRA-SIMPLE CASE OF ONLY ONE REACTANT:

A $\longrightarrow$ PRODUCTS

HERE THE **REACTION RATE** r_A IS THE RATE AT WHICH REACTANT **A** IS USED UP OVER TIME. IT MAY BE EXPRESSED IN MOLES PER SECOND.

IF **A** IS IN SOLUTION, r_A USUALLY REFERS TO THE RATE AT WHICH CONCENTRATION [A] CHANGES, IN MOLES PER LITER PER SECOND, AND IF **A** IS A GAS, r_A MAY REFER EITHER TO CONCENTRATION OR PARTIAL PRESSURE P_A, WHICH AMOUNT TO THE SAME THING.

WELL, THIS ALL SEEMS RATHER ABSTRACT!

FOR EXAMPLE, IN THE LOWER ATMOSPHERE, SUNLIGHT FALLING ON NITROGEN DIOXIDE, NO_2, CAUSES IT TO BREAK INTO NITRIC OXIDE, NO, AND A LOOSE OXYGEN ATOM (CALLED A FREE RADICAL):

$$NO_2 \xrightarrow{\text{LIGHT}} NO + O$$

(THE FREE OXYGEN GOES ON TO BIND WITH O_2 TO FORM OZONE, O_3. OZONE AND THE NITROGEN OXIDES ARE AMONG OUR NASTIER AIR POLLUTANTS.)

CONCRETE ENOUGH FOR YOU?

AT MIDDAY, NO_2 MAKES UP ABOUT 20 PARTS PER BILLION OF THE AIR—20 MOL OF NO_2 PER BILLION MOL OF AIR—OR 20 MOL OF NO_2 IN 24.4×10^9 L OF AIR (AT 25°C). SO MOLAR CONCENTRATION IS $[NO_2] = 20/(24.4 \times 10^9) = 8.2 \times 10^{-10}$ MOL/L. LET'S TAKE AN AIR SAMPLE, AND MEASURE $[NO_2]$ EVERY 40 SECONDS AS IT DECOMPOSES. WE WRITE $[A]_t$ FOR THE CONCENTRATION OF NO_2 AT TIME t.

t (SEC.)	$[A]_t$ ($\times 10^{-10}$ MOL/L)	
0	8.20	$[A]_0$
40	5.80	
80	4.10	$([A]_0)/2$
120	2.90	
160	2.05	$([A]_0)/4$
200	1.45	
240	1.02	$([A]_0)/8$
280	.72	
320	.51	$([A]_0)/16$
360	.36	

THE SPEECH BUBBLES:
- I'M GOING BACK TO THE NINETEENTH CENTURY...
- MUST STARE AT THIS... UNTIL... ITS SECRETS BECOME.... CLEAR...

THE REACTION CERTAINLY SLOWS OVER TIME. IN 10^{10} LITERS OF AIR, **2.4 MOL** ($[A]_0 - [A]_{40}$) WERE USED UP IN THE FIRST 40 SEC., BUT ONLY **0.21 MOL** IN THE 40 SECONDS BETWEEN $t = 280$ AND $t = 320$ ($[A]_{280} - [A]_{320}$).

THE DECLINE HAS A PATTERN: **HALF THE REMAINING REACTANT IS CONSUMED EVERY 80 SECONDS.** AT $t = 80$ SEC., HALF THE NO_2 IS LEFT... AT 160 SEC., A FOURTH REMAINS... AT 240, AN EIGHTH, ETC. WE SAY THE REACTION HAS A **HALF-LIFE,** h, OF 80 SECONDS. DURING **ANY** INTERVAL OF LENGTH h, HALF THE REACTANT IS CONSUMED. IN n HALF LIVES, THEN:

$$[A]_{nh} = (1/2)^n [A]_0$$

n HALF LIVES

143

A SIMPLE MODEL ACCOUNTS FOR THIS BEHAVIOR. START WITH A BIG BUNCH OF MOLECULES OF REACTANT **A**, AND IMAGINE THAT EVERY MOLECULE HAS THE SAME PROBABILITY OF DECOMPOSING. THEN A FIXED FRACTION OF THE WHOLE WILL REACT IN EACH UNIT OF TIME.

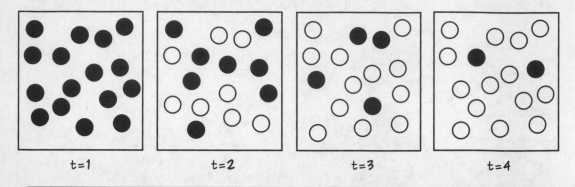

t=1   t=2   t=3   t=4

IN OTHER WORDS, THE REACTION **RATE** (NUMBER OF MOLES OR MOL/L DECOMPOSING PER UNIT TIME) IS PROPORTIONAL TO THE **QUANTITY OF REACTANT** PRESENT (NUMBER OF MOLES OR MOL/L). SO WE CAN WRITE A SECOND FORMULA FOR THE REACTION RATE: AT ANY GIVEN TIME,

$$r_A = -k[A]$$

k IS A CONSTANT CALLED THE **RATE CONSTANT.** BY CONVENTION, k IS ALWAYS A POSITIVE NUMBER, SO THE MINUS SIGN IS NECESSARY TO MAKE r NEGATIVE, MEANING [A] IS DECREASING.

NOTE: MATH-AVERSE READERS MAY SKIP THIS PAGE. OTHERWISE, KEEP READING.

WE CAN EVALUATE k FROM THE DATA. START WITH THE FIRST EQUATION

$$[A]_{nh} = 2^{-n}[A]_O$$

[A] DECREASES EXPONENTIALLY (AS THE EXPONENT OF 2 IN THIS EQUATION). IN PARTICULAR, **[A] NEVER REACHES ZERO.** THEORETICALLY, THE REACTION NEVER ENDS!

h IS AN AWKWARD TIME UNIT—IT VARIES FROM ONE REACTION TO ANOTHER. WE WANT A FIXED UNIT OF TIME, t (DAYS, SECONDS, WHATEVER'S APPROPRIATE). THEN

$$t = nh, \text{ or } n = t/h$$

AND WE CAN WRITE

$$[A]_t = 2^{-t/h}[A]_O$$

TAKING THE NATURAL LOG OF BOTH SIDES,

$$\ln[A]_t = \frac{-1}{h}(\ln 2)t + \ln[A]_O$$

SETTING $k = (1/h)\ln 2$, WE FIND:

$$\ln[A]_t = -kt + \ln[A]_O$$

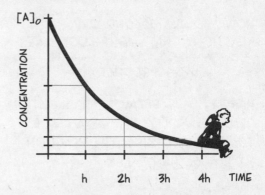

DON'T WORRY! YOU CAN READ ABOUT WORKING WITH LOGARITHMS IN THE APPENDIX!

THAT IS, THE PLOT OF $\ln[A]_t$ AGAINST t IS A **STRAIGHT LINE** WITH SLOPE $-k$. ONE CAN SHOW (USING CALCULUS) THAT THIS IS THE SAME k AS IN $r_A = -k[A]$. IN OUR NO_2 EXAMPLE, THEN,

$$k = (1/80 \text{ SEC})(\ln 2) = (1/80 \text{ SEC})(0.693) = 0.0087 \text{ SEC}^{-1}. \text{ THAT IS,}$$

0.87% OF THE NO_2 GAS IS CONSUMED EVERY SECOND.

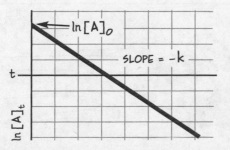

A REACTION WITH $r = -k[A]$ IS CALLED A **FIRST-ORDER REACTION:** IT GOES AS THE FIRST POWER OF A SINGLE CONCENTRATION. YOU CAN CHECK EXPERIMENTALLY IF A REACTION IS FIRST-ORDER BY GRAPHING $\ln[A]_t$ AGAINST t AND SEEING IF IT'S A STRAIGHT LINE. IF SO, THE RATE CONSTANT IS THE NEGATIVE OF THE SLOPE.

Collision Course

HOW ABOUT A SECOND-ORDER
REACTION? THAT MIGHT LOOK LIKE

$$A + B \longrightarrow PRODUCTS$$

HERE $r_A = r_B$ BECAUSE THE REACTION
REMOVES MOLECULES OF A AND B
TOGETHER IN PAIRS. THE REACTION
RATE r IS THEN TAKEN TO BE

$$r = r_A = r_B$$

TO ANALYZE r, THE FIRST THING WE NOTICE IS THAT TWO MOLECULES CAN
COMBINE ONLY IF THEY FIRST **COLLIDE.**

WELL,
DUH!

THIS BRILLIANT OBSERVATION IS THE START OF **COLLISION THEORY.**

HOW OFTEN DO
PARTICLES COLLIDE? IT
DEPENDS ON THEIR
CONCENTRATION (OR
PARTIAL PRESSURE).

IMAGINE THAT A VOLUME OF GAS OR SOLUTION IS DIVIDED INTO COUNTLESS TINY COMPARTMENTS. IF TWO PARTICLES SHARE A COMPARTMENT, WE'LL CALL THAT A COLLISION.

IF [B] IS CONSTANT, THEN CHANGING [A] CHANGES THE NUMBER OF A-B COLLISIONS PROPORTIONALLY. (HERE **A** ARE BLACK AND **B** ARE WHITE.)

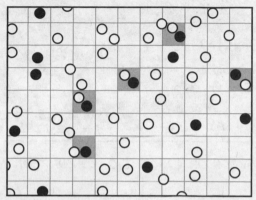

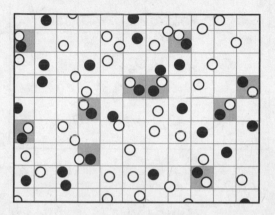

 THE SAME IS TRUE WHEN [B] IS CHANGED, SO THE FREQUENCY OF COLLISIONS MUST BE PROPORTIONAL TO [A][B], OR $P_A P_B$, IF A AND B ARE GASES.

NOT ALL COLLISIONS RESULT IN REACTION. THE ONES THAT DO ARE CALLED **EFFECTIVE.** WE ASSUME THAT THE RATIO OF EFFECTIVE COLLISIONS TO TOTAL COLLISIONS IS CONSTANT (AT A FIXED TEMPERATURE).

SO: REACTION RATE EQUALS RATE OF EFFECTIVE COLLISIONS, WHICH IS PROPORTIONAL TO RATE OF TOTAL COLLISIONS, WHICH IS PROPORTIONAL TO [A][B] OR $P_A P_B$. CONCLUSION:

AMAZING THAT THE LITTLE THINGS EVER MEET AT ALL!

$$r = -k[A][B]$$

k A POSITIVE CONSTANT

WE SAY THE REACTION IS FIRST ORDER IN **A**, FIRST ORDER IN **B**, AND SECOND ORDER OVERALL.

Example

WE'VE ALREADY SEEN THAT IN DAYLIGHT

$$NO_2 \longrightarrow NO + O$$

AND THE MONATOMIC OXYGEN GOES ON TO MAKE OZONE

$$O + O_2 \longrightarrow O_3$$

SO OVERALL

$$NO_2 + O_2 \longrightarrow NO + O_3$$

AT NIGHT, THE REVERSE REACTION TAKES PLACE:

$$NO + O_3 \longrightarrow NO_2 + O_2$$

COUGH   COUGH   COUGH

THIS REACTION HAS RATE r = RATE OF CONSUMPTION OF NO = RATE OF CONSUMPTION OF O_3 AND IS GIVEN BY

$$r = -k[NO][O_3] \qquad k = 1.11 \times 10^7 \ M^{-1}SEC^{-1}$$

A TYPICAL NO CONCENTRATION IS AROUND 24 PPB*, WHICH AS BEFORE GIVES MOLAR CONCENTRATION [NO] AS (24 MOL NO/24.4 $\times$ 10^9 L OF AIR) = 10^{-9} M. [O_3] IS AROUND TWICE THAT, OR 2×10^{-9} M.

A BIT OF CALCULUS PRODUCES THIS PLOT OF THE CONCENTRA-TIONS. THE REACTION GOES QUICKLY: IT'S ESSENTIALLY OVER IN FIVE OR SIX MINUTES.

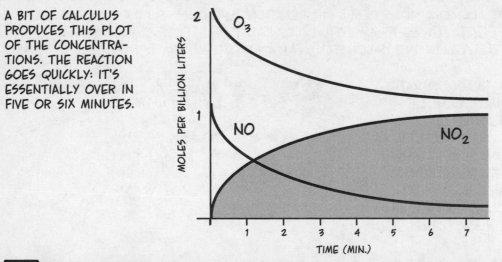

NOTE: THIS GRAPH IS GOOD ONLY FOR AN ISOLATED SAMPLE. TO PREDICT CONCENTRATIONS IN THE ENVIRON-MENT, WE NEED TO KNOW THE RATES OF ALL REACTIONS THAT CONSUME AND PRODUCE NO AND O_3, AS WELL AS HOW MUCH ENTERS THE AIR FROM OUTSIDE SOURCES.

*PARTS PER BILLION

Reactions Up Close

WHY ARE SOME COLLISIONS EFFECTIVE, AND SOME ARE NOT?

ONE REASON IS PARTICLES' RELATIVE **ORIENTATION.** TWO MOLECULES MAY NEED TO PRESENT A CERTAIN "FACE" TO EACH OTHER BEFORE THEY CAN COMBINE. FOR EXAMPLE, WHEN A HIGHLY POLAR MOLECULE OF HCl MEETS ETHENE, CH_2CH_2, A LOT OF ANGLES DON'T WORK.

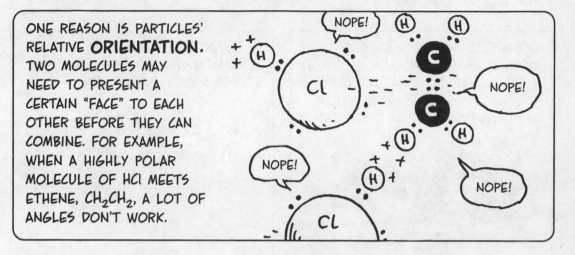

BUT WHEN THE POSITIVE POLE OF HCl MEETS CH_2CH_2'S VERY NEGATIVE DOUBLE BOND, ELECTRONS SHIFT—FIRST, ONE GOES TO HYDROGEN (IT'S CLOSER).

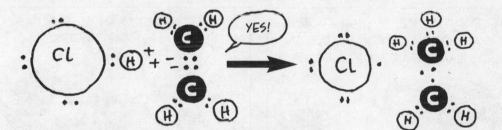

AND THEN ONE GOES TO CHLORINE. THE RESULT IS **CHLOROETHANE,** A TOPICAL ANESTHETIC.

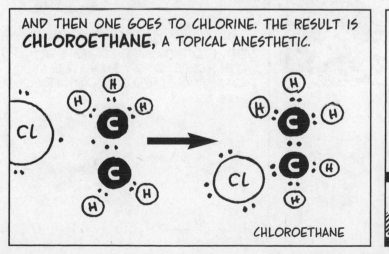

CHLOROETHANE

THE INTERMEDIATE STATE, BEFORE THE CHLORINE IS BONDED, IS CALLED A **TRANSITION STATE.** HERE THE TRANSITION STATE APPEARS ONLY WHEN THE REACTANT MOLECULES ARE ORIENTED PROPERLY.

ANOTHER FACTOR AFFECTING WHETHER COLLISIONS LEAD TO REACTIONS IS HOW FAST THE PARTICLES ARE MOVING.

WHEN FLYING H_2 AND O_2 GAS MOLECULES COLLIDE, FOR INSTANCE, THEIR NEGATIVELY CHARGED ELECTRON CLOUDS REPEL EACH OTHER AND ACTUALLY BECOME DISTORTED.

IF THE KINETIC ENERGY OF THE COLLISION IS TOO LOW, THE MOLECULES SIMPLY BOUNCE AWAY.

BOING

BUT IF INITIAL K.E. IS HIGH ENOUGH TO OVERCOME ELECTRIC REPULSION, THINGS CAN BREAK APART.

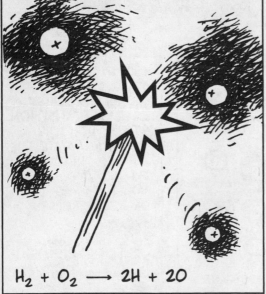

$$H_2 + O_2 \longrightarrow 2H + 2O$$

IF A FREE O MEETS AN H_2, ELECTRIC REPULSION AGAIN DEFORMS THE ELECTRON CLOUDS.

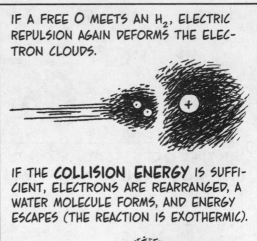

IF THE **COLLISION ENERGY** IS SUFFICIENT, ELECTRONS ARE REARRANGED, A WATER MOLECULE FORMS, AND ENERGY ESCAPES (THE REACTION IS EXOTHERMIC).

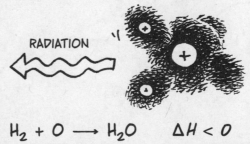

RADIATION

$$H_2 + O \longrightarrow H_2O \qquad \Delta H < 0$$

SO—THE GAS MIXTURE NEEDS SOME EXTRA ENERGY TO GET THE REACTION STARTED: A SPARK OR A FLAME, SAY, TO ENERGIZE SOME PARTICLES.

BUT ONCE IT STARTS, $H_2 + O \rightarrow H_2O$ IS SO **EXOTHERMIC** THAT IT EXCITES THE PARTICLES AROUND IT, AND THE WHOLE REACTION RUSHES FORWARD WITH A SUDDEN, LOUD—

THIS IS ONE REASON WHY CHEMISTS ARE ALWAYS HEATING THINGS... WE HAVE TO SUPPLY THAT INITIAL ENERGY KICK TO GET REACTIONS "OVER THE HUMP."

AND WHY DO **ENDOTHERMIC** REACTIONS KEEP GOING?

SORRY. YOU HAVE TO WAIT UNTIL CHAPTER 10 FOR THE ANSWER TO THAT ONE!

NEARLY EVERY COMBINATION REACTION WORKS THE SAME WAY: IT NEEDS AN ADDED ENERGY PUSH TO BRING THE REACTANTS TOGETHER. THIS BOOST IS CALLED THE **ACTIVATION ENERGY** OF THE REACTION, E_A. IN OTHER WORDS, CHEMICAL REACTIONS ARE NOT JUST LIKE FALLING DOWNHILL!

FALLING DOWNHILL   CHEMICAL REACTION

THE OBVIOUS WAY TO GET A REACTION MOVING FASTER, THEN, IS TO MAKE MORE OF THE PARTICLES EXCEED THE ACTIVATION ENERGY—IN OTHER WORDS, BY **RAISING TEMPERATURE.** THEN A HIGHER FRACTION OF COLLISIONS WILL BE EFFECTIVE.

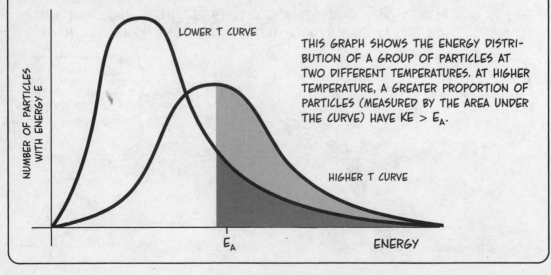

THIS GRAPH SHOWS THE ENERGY DISTRIBUTION OF A GROUP OF PARTICLES AT TWO DIFFERENT TEMPERATURES. AT HIGHER TEMPERATURE, A GREATER PROPORTION OF PARTICLES (MEASURED BY THE AREA UNDER THE CURVE) HAVE $KE > E_A$.

Catalysts, or Raising k

YOU'RE PROBABLY NOT SURPRISED TO HEAR THAT RAISING TEMPERATURE ACCELERATES REACTIONS.* AFTER ALL, WE'VE ALL SEEN IMAGES OF CHEMISTS COOKING THINGS UP. MAYBE WE'VE EVEN TURNED UP THE FLAME A FEW TIMES OURSELVES.

NOW, HOWEVER, WE CAN BE MORE PRECISE. SINCE $r = -k[A][B]$ FOR OUR SECOND-ORDER REACTION, WE CAN SAY THAT BOOSTING TEMPERATURE RAISES k, THE REACTION CONSTANT.

ARE THERE OTHER WAYS TO RAISE k? BASED ON THE PRECEDING DISCUSSION, WE MIGHT WONDER IF IT'S POSSIBLE TO REDUCE A REACTANT'S UNFAVORABLE ORIENTATIONS, OR LOWER THE ACTIVATION ENERGY. THIS IS WHERE **CATALYSTS** COME IN.

*WITHIN LIMITS. WHEN T RISES TOO HIGH, EVERYTHING SHAKES APART, AND THE REACTION IS DISRUPTED.

A **CATALYST** IS A SUBSTANCE THAT SPEEDS UP A REACTION BUT ITSELF EMERGES FROM THE REACTION UNCHANGED.

FOR EXAMPLE, THE **CATALYTIC CONVERTOR** IN A CAR ENGINE SPEEDS THE DETOXIFICATION OF EXHAUST GASES. ONE SUCH REACTION BREAKS CAUSTIC NITRIC OXIDE TO N_2 AND O_2:

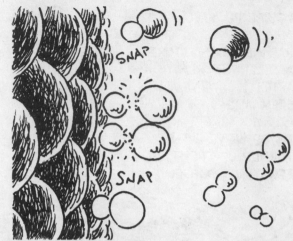

$$2NO \longrightarrow N_2 + O_2$$

IN THE CONVERTOR CHAMBER, PLATINUM, RHODIUM, AND PALLADIUM SCREENS BIND TO THE GAS MOLECULES VIA VARIOUS IMFs.

THE CATALYST BOTH ALIGNS THE **NO** MOLECULES FAVORABLY AND CUTS ACTIVATION ENERGY BY PULLING AGAINST THE N–O BOND—**PROBABLY.** THE EXACT MECHANISM IS UNKNOWN.

CATALYSTS ALSO PROBABLY ENABLED THE **ORIGIN OF LIFE.** THE CHEMICALS OF LIFE (OR PRE-LIFE) WERE TOO BIG AND UNGAINLY TO MAKE PROGRESS BY RANDOM COMBINATION... BUT IF (AS SEVERAL THEORIES SUGGEST) THEY WERE ANCHORED AT ONE END TO A CHARGED SURFACE, SUCH AS CLAY ON THE OCEAN FLOOR, THEY WOULD BE MUCH MORE LIKELY TO ENGAGE IN "GOOD" REACTIONS!

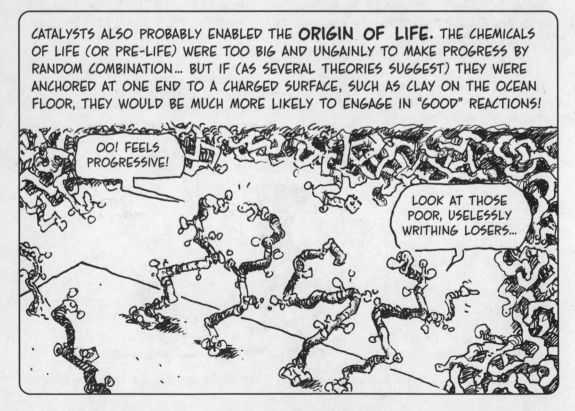

Higher-order Reactions, Maybe

WE SAW THAT

$$A + B \longrightarrow \text{PRODUCTS}$$

IS A SECOND-ORDER REACTION WITH RATE $r = -k[A][B]$. THIS, BY THE WAY, INCLUDES THE SPECIAL CASE WHEN **A AND B ARE THE SAME.** THE REACTION

$$A + A \longrightarrow \text{PRODUCTS}$$

HAS A RATE $-k[A]^2$.

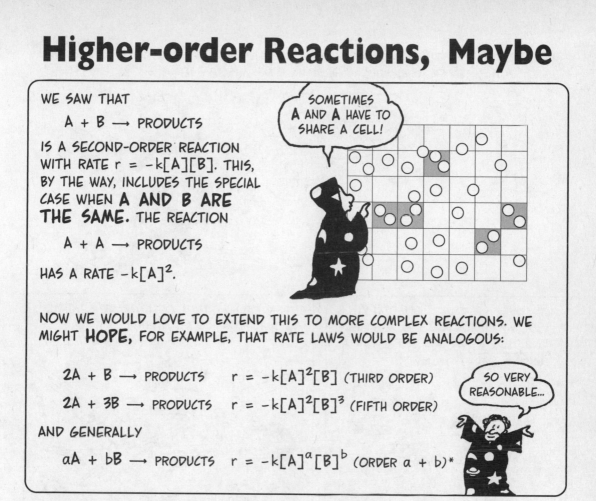

SOMETIMES A AND A HAVE TO SHARE A CELL!

NOW WE WOULD LOVE TO EXTEND THIS TO MORE COMPLEX REACTIONS. WE MIGHT **HOPE,** FOR EXAMPLE, THAT RATE LAWS WOULD BE ANALOGOUS:

$$2A + B \longrightarrow \text{PRODUCTS} \qquad r = -k[A]^2[B] \text{ (THIRD ORDER)}$$

$$2A + 3B \longrightarrow \text{PRODUCTS} \qquad r = -k[A]^2[B]^3 \text{ (FIFTH ORDER)}$$

AND GENERALLY

$$aA + bB \longrightarrow \text{PRODUCTS} \qquad r = -k[A]^a[B]^b \text{ (ORDER } a + b\text{)*}$$

SO VERY REASONABLE...

WE WOULD LOVE TO SAY IT, READER, BUT UNFORTUNATELY WE CAN'T, BECAUSE IT'S

false.

RATES OF REAL-LIFE REACTIONS CAN'T BE PREDICTED FROM THEORY, BUT **MUST BE MEASURED EXPERIMENTALLY.**

WELL, SOMETIMES IT'S TRUE...

*WE HAVE TO BE A LITTLE CAREFUL ABOUT WHAT WE MEAN BY r. IT'S THE RATE AT WHICH $aA + bB$ IS CONSUMED. THAT IS, $r = (1/a)r_A = (1/b)r_B$.

IN FACT, EVEN THE REACTION (A + B → PRODUCTS) SOMETIMES DOESN'T BEHAVE AS WE CLAIMED. YES, READER, MUCH OF THE FIRST HALF OF THIS CHAPTER IS SIMPLY **UNTRUE!**

USEFUL? CERTAINLY! CONCEPTUALLY VALID? SORT OF...

BUT TRUE?

WE COVERTLY MADE A **SIMPLIFYING ASSUMPTION**, YOU SEE, BY IMAGINING THAT REACTIONS HAPPEN IN A **SINGLE STEP**.

YES... SO... WHY NOT?

BUT IN REALITY THEY OFTEN TAKE SEVERAL STEPS TO COMPLETE... SORRY!

FOR INSTANCE, WHEN WE WRITE 2A + B, ARE WE REALLY TO IMAGINE THREE PARTICLES COLLIDING AT ONCE? NOT LIKELY... MORE PROBABLY, **A** MEETS **B** TO FORM **AB**, THEN ANOTHER **A** COMES ALONG...

I'VE BEEN LIED TO...

ONE-STEP REACTIONS ARE CALLED **ELEMENTARY**... AND IT **IS** TRUE THAT AN ELEMENTARY REACTION aA + bB → PRODUCTS HAS A REACTION RATE OF

$$r = -k[A]^a[B]^b.$$

WELL, THAT'S SOMETHING, ANYWAY...

IN A MULTI-STEP REACTION, INTERMEDIATE STEPS ARE OFTEN UNCLEAR... THINGS GO BY TOO FAST TO OBSERVE. BUT THIS MUCH IS TRUE: THE **SLOWEST INTERMEDIATE REACTION RATE** DETERMINES THE OVERALL RATE.

TO SEE THIS, IMAGINE A WASHER-DRYER COMBO THAT PROCESSES A LOAD OF DIRTY CLOTHES IN EXACTLY **24 HOURS.** LET'S LIFT THE LID AND SEE HOW IT WORKS...

WASHING, IT SEEMS, IS DONE MANUALLY BY ILL-TRAINED, UNCOOPERATIVE WEASELS WHO TAKE **23.999 HOURS** TO DO A LOAD. THE DRYER IS A NUCLEAR BLAST FURNACE THAT CRISPS YOUR CLOTHES IN A MILLISECOND.

PROCESS 1: RATE = ONE LOAD/DAY   PROCESS 2: RATE = 86.4 MILLION LOADS/DAY

OVERALL PROCESS: RATE = ONE LOAD/DAY

NOW IS IT CLEAR THAT THE OVERALL RATE IS THE RATE OF THE SLOWEST STEP? WHEN THE WEASELS ARE DONE, THE "REACTION" IS ALL BUT OVER!

CHEMICAL EXAMPLE: IODIDE ION REDUCES PEROXYDISULFATE

$$S_2O_8{}^{2-} + 2I^- \longrightarrow 2SO_4{}^{2-} + I_2$$

LOOKS THIRD-ORDER, BUT EXPERIMENT SAYS SECOND-ORDER, WITH

$$r = -k[S_2O_8{}^{2-}][I^-]$$

BLASTED WEASELS!

CHEMISTS PROPOSE TWO ELEMENTARY STEPS:

$$S_2O_8{}^{2-} + I^- \longrightarrow 2SO_4{}^{2-} + I^+$$
$$I^+ + I^- \longrightarrow I_2$$

THE FIRST REACTION'S THEORETICAL RATE

$$r = -k[S_2O_8{}^{2-}][I^-]$$

MATCHES THE OBSERVED RATE OF THE OVERALL REACTION. THE SECOND REACTION PRESUMABLY HAPPENS VERY FAST.

Equilibrium...

IS A STATE OF **DYNAMIC BALANCE.** IN NATURE, WE OFTEN FIND TWO PROCESSES THAT **UNDO** EACH OTHER—EVAPORATION AND CONDENSATION, FOR INSTANCE. WHEN THE PROCESSES UNDO EACH OTHER AT THE **SAME RATE,** NOTHING APPEARS TO BE CHANGING. THAT'S **EQUILIBRIUM.**

IF I SOIL MY CLOTHES AT THE SAME RATE THEY'RE WASHED AND DRIED, I ALWAYS HAVE THE SAME AMOUNT OF CLEAN CLOTHES.

I'M IN EQUILIBRIUM WITH WEASELS...

FILTHY PIG...

MANY CHEMICAL REACTIONS ARE **REVERSIBLE.**

$$aA + bB \rightleftharpoons cC + dD$$

REACTANTS **A** AND **B** COMBINE TO MAKE **C** AND **D**... BUT IF EVERYTHING REMAINS MIXED TOGETHER, **C** AND **D** CAN FIND EACH OTHER TO MAKE **A** AND **B.**

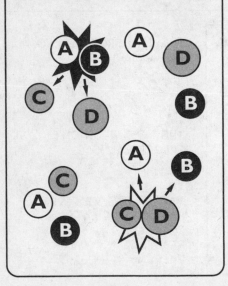

WE SAW AN EXAMPLE IN CHAPTER 4:

$$CaCO_3(s) \rightleftharpoons CaO(s) + CO_2\uparrow$$

LIMESTONE WAS COOKED TO FORM QUICKLIME AND CARBON DIOXIDE GAS. LATER, THE WHITE-WASH MADE FROM CaO REACTED WITH CO_2 FROM THE ATMOSPHERE TO MAKE $CaCO_3$ AGAIN.

CHALKY!

IF THE CO_2 HAD NOT BEEN ALLOWED TO ESCAPE IN THE ORIGINAL REACTION (I.E., IF THE REACTION HAD OCCURRED IN A CLOSED VESSEL), SOME OF THE GAS WOULD HAVE RECOMBINED THEN AND THERE.

NOW IMAGINE A REACTION VESSEL CONTAINING THE REACTANTS **A** AND **B**.

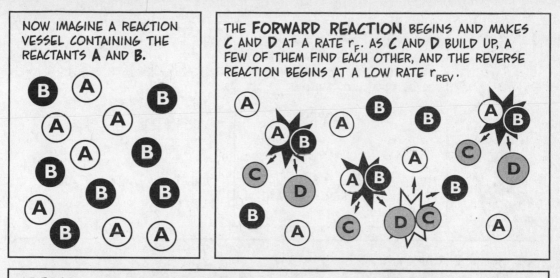

THE **FORWARD REACTION** BEGINS AND MAKES C AND D AT A RATE r_F. AS C AND D BUILD UP, A FEW OF THEM FIND EACH OTHER, AND THE REVERSE REACTION BEGINS AT A LOW RATE r_{REV}.

AT FIRST, $r_F > r_{REV}$, AND THE REACTION "GOES TO THE RIGHT." **A** AND **B** ARE CONSUMED FASTER THAN THEY ARE REPLENISHED, AND C AND D BUILD UP FASTER THAN THEY ARE CONSUMED.

IN OTHER WORDS, AS LONG AS $r_F > r_{REV}$, [A] AND [B] FALL AND [C] AND [D] RISE.

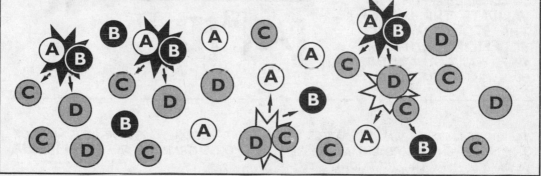

BUT RATES ARE PROPORTIONAL TO (POWERS OF) CONCENTRATIONS. SO AS LONG AS $r_F > r_{REV}$, r_F MUST FALL AND r_{REV} MUST RISE. **THE REACTION CONTINUES UNTIL**

$$r_F = r_{REV}.$$

AT THIS POINT, EACH SUBSTANCE IS BEING CONSUMED AT THE SAME RATE IT IS BEING REPLENISHED. THE CONCENTRATIONS [A], [B], [C], AND [D] NO LONGER CHANGE. THE REACTION HAS REACHED **EQUILIBRIUM.**

A LOT IS GOING ON, BUT VERY QUIETLY!

AND A LITTLE MORE MATH...

NOW WE MAKE AN UNWARRANTED ASSUMPTION: SUPPOSE THE REACTION ORDERS ARE GIVEN BY THE STOICHIOMETRIC COEFFICIENTS a, b, c, AND d. THAT IS:

$$r_F = -k_F[A]^a[B]^b$$

$$r_{REV} = -k_{REV}[C]^c[D]^d$$

(HERE k_F AND k_{REV} ARE THE FORWARD AND REVERSE RATE CONSTANTS.)

AT EQUILIBRIUM, THEN, THE RATES ARE EQUAL:

$$k_F[A]^a[B]^b = k_{REV}[C]^c[D]^d$$

REARRANGING,

$$\frac{[C]^c[D]^d}{[A]^a[B]^b} = \frac{k_F}{k_{REV}} = K,$$

WHERE K IS A **CONSTANT.**

BUT WHAT IF OUR ASSUMPTION IS WRONG, AND THOSE ARE NOT THE REAL RATES? NO PROBLEM! BY SOME MIRACLE, ALL INTERMEDIATE STEPS CAN BE SHOWN TO COMBINE PERFECTLY TO **VALIDATE THE USE OF THE STOICHIOMETRIC COEFFICIENTS.** THAT IS, THERE REALLY IS A CONSTANT K, SUCH THAT AT EQUILIBRIUM:

$$\frac{[C]^c[D]^d}{[A]^a[B]^b} = K$$

TO PUT IT ANOTHER WAY, NO MATTER WHERE THE REACTION STARTS OR HOW MUCH OF ANY INGREDIENT IS PRESENT AT ANY TIME, THE CONCENTRATIONS **AT EQUILIBRIUM** ALWAYS SATISFY THE EQUATION:

$$\frac{[C]^c[D]^d}{[A]^a[B]^b} = K$$

THIS FACT IS CALLED THE **law of mass action,** AND K IS THE REACTION'S **equilibrium constant.**

GEE... THREE TIMES ON ONE PAGE... THINK THAT'S ENOUGH?

NO!

$$\frac{[C]^c[D]^d}{[A]^a[B]^b} = K$$

Example: Ionization of water

CONSIDER $H_2O \rightleftharpoons H^+ + OH^-$. WATER MOLECULES OCCASIONALLY BREAK APART, AND H^+ AND OH^- REACH AN EQUILIBRIUM CONCENTRATION.

POP

PRECISE MEASUREMENT OF PURE WATER AT 25°C SHOWS $[H^+]$ AND $[OH^-]$ TO BE ALMOST EXACTLY 10^{-7} M – NOT MUCH!

THREE-EYED FREAK!

H^+ IONS ALWAYS ATTACH THEMSELVES TO A WATER MOLECULE TO MAKE H_3O^+.

WE PLUG IN THOSE VALUES AND CALCULATE THE EQUILIBRIUM CONSTANT.

$$K = \frac{[H^+][OH^-]}{[H_2O]} = \frac{(10^{-7})(10^{-7})}{[H_2O]} = \frac{10^{-14}}{[H_2O]}$$

IT'S SO SMALL...

TRUE... BUT EVEN AT 10^{-7} M, THERE ARE ABOUT 60,000,000,000,000,000 OF EACH ION IN A LITER!

WHAT'S $[H_2O]$? BEFORE DISSOCIATION, IT'S 55.6 M. (1 L OF WATER WEIGHS 1000g; 1 MOL WATER WEIGHS 18 g; 1000/18 = 55.6.) AFTER DISSOCIATION, IT'S

55.6 – 0.0000001

BARELY DIFFERENT. SO WE CAN SAY

$$K = \frac{10^{-14}}{55.6}$$

AND WE USE THIS HOW?

NOW SUPPOSE 0.1 MOL OF **HYDRO-CHLORIC ACID**, HCl, DISSOLVES IN A LITER OF WATER. HCl, A POLAR MOLECULE, ALMOST COMPLETELY DISSOCIATES INTO H^+ AND Cl^- IONS. SUDDENLY, $[H^+]$ RISES TO 0.1 M. THEN WHAT?

GRR-RR!

THEY DON'T CALL IT A CONSTANT FOR NOTHING! WE IMMEDIATELY WRITE

$$10^{-14} = 55.6K = [H^+][OH^-]$$
$$= (0.1)[OH^-]$$

SOLVING FOR $[OH^-]$,

$$[OH^-] = 10^{-13}$$

THAT IS, THE ADDED H^+ IONS GOBBLED UP EXACTLY ENOUGH OH^- IONS TO MAINTAIN THE PRODUCT $[H^+][OH^-]$ AT A CONSTANT 10^{-14}.

Le Chatelier's Principle

YOU CAN THINK OF EQUILIBRIUM AS A BALANCED SEESAW WITH REACTANTS ON ONE SIDE AND PRODUCTS ON THE OTHER. IN THE LAST EXAMPLE, H_2O WAS ON THE LEFT, OH^- AND H^+ ON THE RIGHT.

IN THAT EXAMPLE, THE EQUILIBRIUM WAS DISTURBED BY ADDING H^+ TO THE RIGHT SIDE. WHAT HAPPENS THEN?

THE FRENCH CHEMIST **HENRY LE CHATELIER** HAS LEFT US A GENERAL PRINCIPLE FOR ANALYZING WHAT HAPPENS WHEN CHEMICAL EQUILIBRIUM IS DISTURBED.

When an external stress is applied to a system at equilibrium, the process evolves in such a way as to reduce the stress.

FOR EXAMPLE, IF $aA + bB \rightleftharpoons cC + dD$ IS IN EQUILIBRIUM, THEN ADDING REACTANT **A** DRIVES THE REACTION TO THE RIGHT—CONSUMING MORE **A.**

IN OUR EXAMPLE, ADDING LOADS OF H^+ TO THE RIGHT-HAND SIDE OF $H_2O \rightleftharpoons H^+ + OH^-$ DROVE THE REACTION TO THE LEFT.

$[OH^-]$ FELL SHARPLY, AND EVERY OH^- ION THAT DISAPPEARED TOOK AN H^+ WITH IT, THEREBY LOWERING $[H^+]$.

THO' QUITE A LOT OF H^+ REMAINED IN THE END!

LE CHATELIER VERY CLEVERLY APPLIED HIS OWN PRINCIPLE TO THE SYNTHESIS OF **AMMONIA**, NH_3, A KEY INGREDIENT OF COUNTLESS PRODUCTS, FROM FERTILIZER TO EXPLOSIVES.

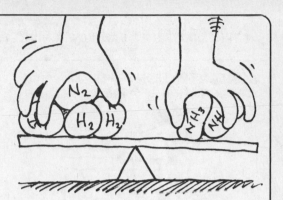

$$N_2(g) + 3H_2(g) \rightleftharpoons 2NH_3(g)$$

INCREASING **PRESSURE**, SAID HIS PRINCIPLE, WILL DRIVE THE REACTION IN THE DIRECTION THAT **REDUCES PRESSURE.**

THERE ARE FOUR MOLES OF GAS ON THE LEFT, BUT ONLY TWO ON THE RIGHT. BY THE GAS LAW, PRESSURE IS DIRECTLY PROPORTIONAL TO THE NUMBER OF MOLES. SO PRESSURE IS RELIEVED WHEN THE REACTION GOES IN THE DIRECTION OF **FEWER MOLES,** THAT IS, TO THE RIGHT.

IN 1901, LE CHATELIER ATTEMPTED THE SYNTHESIS AT A PRESSURE OF **200** atm IN A STEEL "BOMB" HEATED TO **600°C.** UNFORTUNATELY, AN AIR LEAK CAUSED THE BOMB TO EXPLODE...

...AND THE CHEMIST GAVE UP THIS FERTILE LINE OF INVESTIGATION.

I CAN'T TAKE THE PRESSURE...

FIVE YEARS LATER, THE GERMAN **FRITZ HABER** SUCCEEDED WHERE LE CHATELIER HAD FAILED, AND EVER SINCE, AMMONIA SYNTHESIS HAS BEEN KNOWN AS THE

Haber process.

"I LET THE DISCOVERY OF THE AMMONIA SYNTHESIS SLIP THROUGH MY HANDS. IT WAS THE GREATEST BLUNDER OF MY SCIENTIFIC CAREER."
—LE CHATELIER

IN THIS CHAPTER, WE SAW HOW A NUMBER OF FACTORS AFFECTED REACTION RATES:

CONCENTRATION: RAISING CONCENTRATION UPS THE RATE.

TEMPERATURE: RAISING TEMPERATURE UPS THE RATE.

ACTIVATION ENERGY: LOWERING IT, BY MEANS OF A CATALYST, UPS THE RATE.

WE ALSO SAW HOW A BUILDUP OF REACTION PRODUCTS COULD START A REVERSE REACTION THAT OVERTAKES THE FORWARD REACTION AT **EQUILIBRIUM.**

IN THE NEXT CHAPTER, WE'LL EXPLORE SOME GREAT USES OF THE CONCEPT—AND THE CONSTANT—OF EQUILIBRIUM, AND IN THE CHAPTER AFTER THAT, WE'LL DIG DEEP AND DISCOVER WHAT EQUILIBRIUM **REALLY MEANS.**

Chapter 9
Acid Basics

ACIDS, SOUR AND AGGRESSIVE, ARE EVERYWHERE: IN SALAD DRESSING, RAINWATER, CAR BATTERIES, SOFT DRINKS, AND YOUR STOMACH. THEY CAN BURN, CORRODE, DIGEST, OR ADD A PLEASANT TANG TO FOOD AND DRINK...

BASES, BITTER AND SLIPPERY, MAY BE LESS FAMILIAR, BUT ARE EXACTLY AS COMMON AS ACIDS. YOU'LL FIND THEM IN BEER, BUFFERIN, SOAP, BAKING SODA, AND DRAIN CLEANERS...

ACIDS AND BASES ARE SOMETIMES USEFUL, OFTEN HARMFUL, AND ALWAYS A GREAT OPPORTUNITY TO PLAY WITH EQUILIBRIUM CONSTANTS!

ACIDS AND BASES ARE INTIMATELY CONNECTED VIA PROTONS, I.E., HYDROGEN IONS, H⁺.

AN **ACID** IS ANY SUBSTANCE THAT THROWS OFF PROTONS. THE STRONGER THE ACID, THE MORE EASILY IT SHEDS H⁺.

WEAK ACID

STRONG ACID

SINCE NAKED PROTONS ARE WILD, AGGRESSIVE CREATURES, STRONG ACIDS ARE VERY REACTIVE.

AAARGH...

A **BASE** IS ANY SUBSTANCE THAT TAKES UP PROTONS. BASES GENERALLY HAVE AN EXPOSED ELECTRON PAIR WHERE A PROTON CAN NESTLE.

YOO-HOO!

THE STRONGER THE BASE, THE MORE STRONGLY IT WANTS TO BOND TO A PROTON.

ER...

WEAK BASE

STRONG BASE

AS YOU CAN SEE, AN ACID IS JUST A PROTON ATTACHED TO A BASE! AN ACID AND BASE PAIRED IN THIS WAY ARE CALLED **CONJUGATE** TO EACH OTHER.

BY DEFINITION, THE STRONGER AN ACID, THE WEAKER ITS CONJUGATE BASE, AND VICE VERSA.

STRONG ACID, WEAK CONJUGATE BASE, LOOSE PROTON

WEAK ACID, STRONG CONJUGATE BASE, TIGHTLY BOUND PROTON

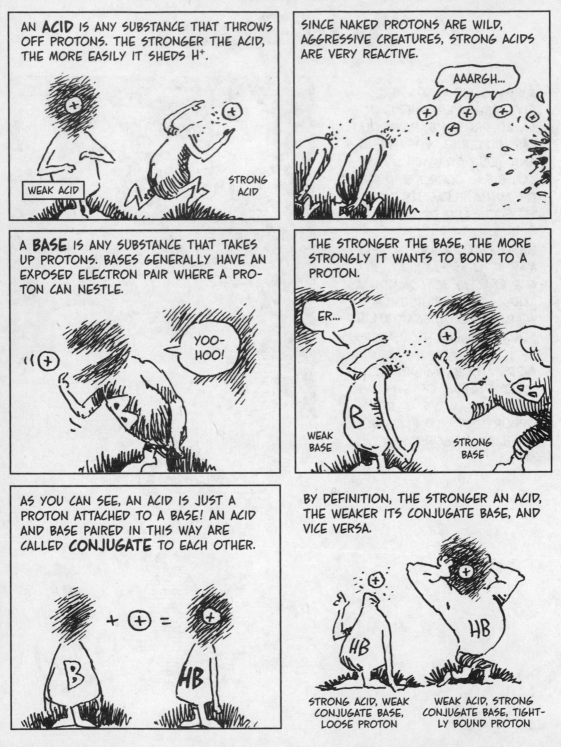

166

SOME CONJUGATE ACID-BASE PAIRS:

ACIDS, STRONGEST TO WEAKEST	BASES, WEAKEST TO STRONGEST
SULFURIC, H_2SO_4	BISULFATE, HSO_4^-
HYDROIODIC, HI	IODIDE, I^-
HYDROBROMIC, HBr	BROMIDE, Br^-
HYDROCHLORIC, HCl	CHLORIDE, Cl^-
NITRIC HNO_3	NITRATE, NO_3
HYDRONIUM, H_3O^+	WATER H_2O
BISULFATE, HSO_4^-	SULFATE, SO_4^{2-}
SULFUROUS, H_2SO_3	BISULFITE, HSO_3^-
PHOSPHORIC, H_3PO_4	$H_2PO_4^-$
HYDROFLUORIC, HF	FLUORIDE, F^-
NITROUS HNO_2	NITRITE NO_2^-
ACETIC (VINEGAR), CH_3CO_2H	ACETATE, $CH_3CO_2^-$
CARBONIC H_2CO_3	BICARBONATE, HCO_3^-
AMMONIUM NH_4^+	AMMONIA NH_3
HYDROCYANIC, HCN	CYANIDE, CN^-
BICARBONATE, HCO_3^-	CARBONATE, CO_3^{2-}
WATER, H_2O	HYDROXIDE, OH^-

NOTE: BOTH ACIDS AND BASES CAN BE EITHER CHARGED OR NEUTRAL.

Acids and Bases in Water

NOW WE WOULD LIKE A **NUMERICAL MEASURE** OF AN ACID'S STRENGTH. THIS IS EASIEST FOR ACIDS DISSOLVED IN WATER. (MOST ACIDS WE ENCOUNTER IN THE WORLD AND IN THE LAB ARE WATER SOLUBLE.)

IMPORTANT SAFETY NOTE: ALWAYS ADD **ACID TO WATER**, NEVER VICE VERSA. WEAR GLOVES WHEN HANDLING STRONG ACIDS.

WHEN A **STRONG** ACID DISSOLVES IN WATER, THE ACID COMPLETELY **IONIZES**, OR DISSOCIATES. HYDROCHLORIC ACID, FOR EXAMPLE, DOES THIS:

$$HCl \longrightarrow H^+ + Cl^-$$

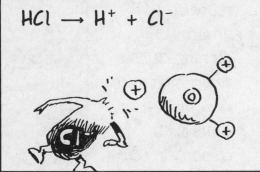

BUT THAT PROTON CAN'T FLOAT AROUND FREELY: ITS CHARGE SOON DRAWS A CLUSTER OF WATER MOLECULES.

COZY...

FOR CONVENIENCE, WE ASSIGN IT TO ONE OF THESE H_2O MOLECULES, AND WE CALL THE CLUSTER A **HYDRONIUM** ION, H_3O^+. IN SHORT,

$$HCl + H_2O \longrightarrow H_3O^+ + Cl^-$$

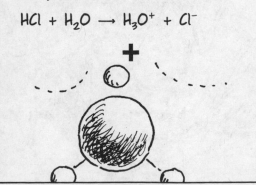

THAT IS, ADDING STRONG ACID TO WATER RAISES THE CONCENTRATION OF H_3O^+. H_3O^+ IS A STRONG ACID, AND THE HIGHER ITS CONCENTRATION, THE MORE ACIDIC THE SOLUTION.

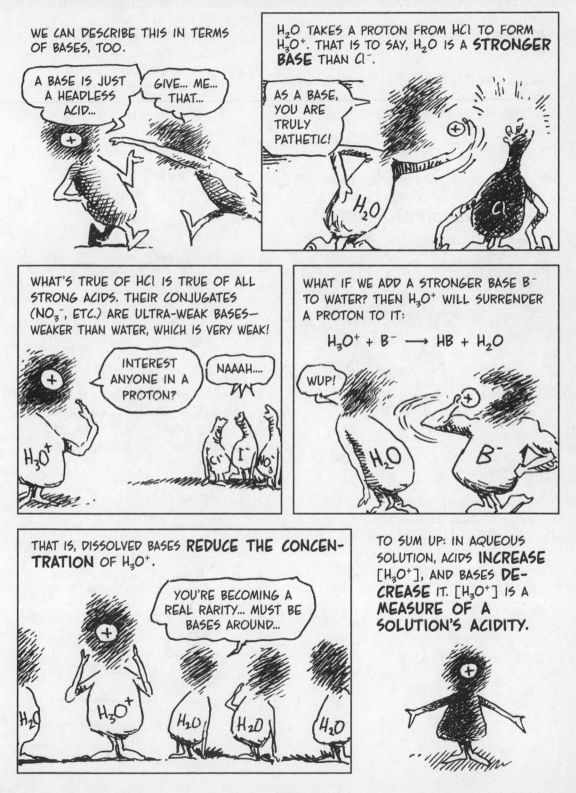

pH

HOW HIGH IS $[H_3O^+]$? LET'S REVIEW THE DISCUSSION ON PAGE 161 IN CHAPTER 8. WATER ALWAYS IONIZES ITSELF A LITTLE:

$$H_2O + H_2O \rightleftharpoons H_3O^+ + OH^-$$

AT EQUILIBRIUM, IN PURE WATER AT 25°C, THE MOLAR CONCENTRATIONS OF H_3O^+ AND OH^- ARE BOTH 1.0×10^{-7} M.

OH, BOY! HERE COMES OUR FIRST EQUILIBRIUM CONSTANT!

THE EQUILIBRIUM CONSTANT FOR THIS REACTION IS

$$K_{eq} = \frac{[H_3O^+][OH^-]}{[H_2O]^2}$$

BUT THE DENOMINATOR IS CONSTANT, OR NEARLY SO. ONLY ABOUT ONE WATER MOLECULE IN 556,000,000 IONIZES! THEREFORE THE NUMERATOR IS A CONSTANT TOO. WE CALL IT THE **WATER CONSTANT.**

$$K_w = [H_3O^+][OH^-]$$
$$= (10^{-7})(10^{-7})$$
$$= 10^{-14}.$$

A STRONG ACID GIVES ALL ITS PROTONS TO WATER TO MAKE H_3O^+. FOR INSTANCE, A 1 M SOLUTION OF HNO_3 HAS

$$[H_3O^+] = 1 M = 10^0 M$$

SO

$$[OH^-] \text{ DROPS TO } K_w/[H_3O^+]$$
$$= 10^{-14}$$

SO PRECISE, IT'S LIKE MAGIC!

ON THE OTHER HAND, A BASIC COMPOUND LIKE NaOH DISSOCIATES FULLY IN WATER AND RAISES $[OH^-]$. $[H_3O^+]$ FALLS ACCORDINGLY. A 1 M SOLUTION OF NaOH HAS

$$[OH^-] = 1$$
$$[H_3O^+] = 10^{-14}.$$

REALLY TINY!

WELL... THAT'S STILL ABOUT 6,000,000,000 H_3O^+ IONS PER LITER!

FOR MOST PRACTICAL PURPOSES, THEN, $[H_3O^+]$ FLUCTUATES BETWEEN 1 AND 10^{-14}.

NOW WHEN CHEMISTS SEE 10^x, THEY OFTEN FIND IT SIMPLER TO TALK ABOUT x, THE LOGARITHM. THEY DEFINE

$$pH = -\log [H_3O^+]$$

pH STANDS FOR POWER OF HYDROGEN. pH RANGES APPROXIMATELY FROM 0 TO 14. THE LOWER THE pH, THE MORE ACIDIC THE SOLUTION. FOR INSTANCE, A 0.01 M SOLUTION OF HCl HAS $[H_3O^+]$ = .01 = 10^{-2}, SO pH = 2.

WHEN DEALING WITH BASES, IT CAN BE MORE CONVENIENT TO USE pOH. THIS IS DEFINED AS

$$pOH = -\log[OH^-]$$

AND WE HAVE

$$pH + pOH = 14$$

pH	SUBSTANCE
0	5% SULFURIC ACID
1	STOMACH ACID
2	LEMONS VINEGAR
3	APPLES, GRAPEFRUIT COCA-COLA, ORANGES
4	TOMATOES, ACIDIFIED LAKES
5	COFFEE BREAD POTATOES
6	NATURAL RIVERS
7	MILK PURE WATER, SALIVA TEARS, BLOOD
8	SEA WATER
9	BAKING SODA
10	WATER IN MONO LAKE
11	MILK OF MAGNESIA
12	
13	LIME WATER
14	LYE, 4% SODIUM HYDROXIDE

Weak Ionization

IN WATER, STRONG ACIDS IONIZE... WELL... STRONGLY. WHEN HCl DISSOLVES, IT RELEASES VIRTUALLY ALL ITS HYDROGEN AS H^+, AND pH IS GIVEN DIRECTLY BY HOW MUCH HCl IS IN SOLUTION.

BUT A COMPLICATION ARISES WITH H_2SO_4, A STRONG ACID WITH TWO PROTONS TO GIVE. ONLY THE FIRST ONE IONIZES COMPLETELY:

$$H_2SO_4 + H_2O \longrightarrow H_3O^+ + HSO_4^-$$

YES!

NO!

HSO_4^- IS A WEAKER ACID, WHICH PARTS WITH ITS PROTON LESS WILLINGLY.

HOW DO WE SPECIFY THE "ACIDITY" OF WEAK ACIDS? THESE ACIDS IONIZE ONLY PARTWAY IN WATER. THAT IS, IF HB IS ANY WEAK ACID IN AQUEOUS SOLUTION, IT SOMETIMES HANDS OFF ITS H^+ TO H_2O, AND SOMETIMES THE PROTON COMES BACK:

$$HB + H_2O \rightleftharpoons H_3O^+ + B^-$$

OH, BOY! I FEEL AN EQUILIBRIUM CONSTANT COMING ON!!!

THE REACTION'S EQUILIBRIUM CONSTANT EXPRESSES THE EXTENT OF IONIZATION:

$$\frac{[H_3O^+][B^-]}{[HB][H_2O]}$$

AS USUAL, $[H_2O]$ IS CONSTANT, SO WE CAN REMOVE IT FROM THE EXPRESSION. THEN THE **ACID IONIZATION CONSTANT** K_a IS DEFINED BY

THE MORE IONIZED THE ACID, THE BIGGER I AM!

$$K_a = \frac{[H^+][B^-]}{[HB]}$$

WITH H_3O^+ ABBREVIATED AS H^+...

HERE ARE K_a VALUES FOR A FEW WEAK ACIDS. A HIGH VALUE FOR K_a MEANS A LARGE NUMERATOR, THAT IS, A LOT OF IONS RELATIVE TO THE NON-IONIZED SPECIES IN THE DENOMINATOR. THAT IS, HIGHER K_a MEANS STRONGER ACID.

		K_{a1}	K_{a2}
ACETIC	CH_3CO_2H	1.75×10^{-5}	
CARBONIC	H_2CO_3	4.45×10^{-7}	4.7×10^{-11}
FORMIC	HCO_2H	1.77×10^{-4}	
HYDROFLUORIC	HF	7.0×10^{-4}	
HYPOCHLOROUS	$HOCl$	3.0×10^{-8}	
NITROUS	HNO_2	4.6×10^{-4}	
SULFURIC	H_2SO_4	STRONG	1.20×10^{-2}
SULFUROUS	H_2SO_3	1.72×10^{-2}	6.43×10^{-8}

AND K_{a2} MEANS:

ACIDS THAT SHED MORE THAN ONE PROTON WILL HAVE MORE THAN ONE IONIZATION CONSTANT. FOR EXAMPLE, H_2CO_3, WHICH CAN SHED TWO PROTONS, HAS K_{a1} FOR

$$H_2CO_3 \rightleftharpoons H^+ + HCO_3$$

AND K_{a2} FOR

$$HCO_3^- \rightleftharpoons H^+ + CO_3^{2-}$$

THE FIRST PROTON COMES OFF MORE EASILY THAN THE SECOND!

NOTE ALSO: IN WATER SOLUTION, SOME METAL IONS CAN ACT AS ACIDS. BY GRABBING OH^- FROM WATER, THEY GENERATE H_3O^+. Fe^{3+} IS AN EXAMPLE:

BRING ME THE WORLD'S BIGGEST BOX OF BAKING SODA!

$$Fe^{3+} + 2H_2O \rightleftharpoons FeOH^{2+} + H_3O^+$$
$$FeOH^{2+} + 2H_2O \rightleftharpoons Fe(OH)_2^+ + H_3O^+$$
$$Fe(OH)_2^+ + 2H_2O \rightleftharpoons Fe(OH)_3 + H_3O^+$$

ACID MINE DRAINAGE CONTAINS Fe^{3+}. WHEN IT ENTERS A RIVER WITH HIGHER pH, IT PRECIPITATES OUT AS AN UGLY SLIME CALLED "YELLOW BOY."

Example

K_a CAN BE USED TO FIND THE pH OF A WEAK ACID SOLUTION.

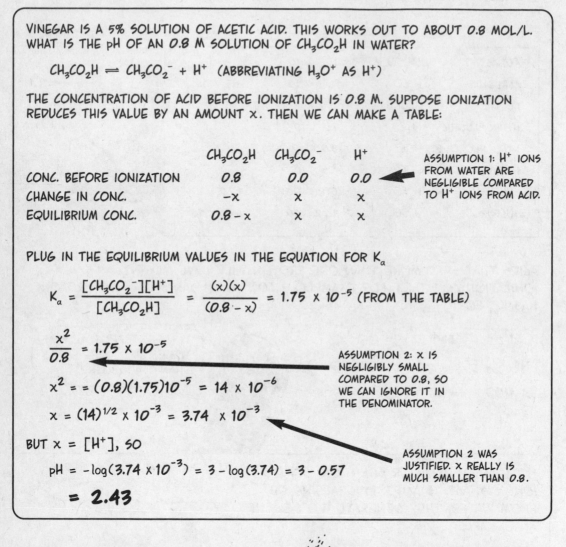

VINEGAR IS A 5% SOLUTION OF ACETIC ACID. THIS WORKS OUT TO ABOUT 0.8 MOL/L. WHAT IS THE pH OF AN 0.8 M SOLUTION OF CH_3CO_2H IN WATER?

$$CH_3CO_2H \rightleftharpoons CH_3CO_2^- + H^+ \text{ (ABBREVIATING } H_3O^+ \text{ AS } H^+\text{)}$$

THE CONCENTRATION OF ACID BEFORE IONIZATION IS 0.8 M. SUPPOSE IONIZATION REDUCES THIS VALUE BY AN AMOUNT x. THEN WE CAN MAKE A TABLE:

	CH_3CO_2H	$CH_3CO_2^-$	H^+
CONC. BEFORE IONIZATION	0.8	0.0	0.0
CHANGE IN CONC.	$-x$	x	x
EQUILIBRIUM CONC.	$0.8 - x$	x	x

ASSUMPTION 1: H^+ IONS FROM WATER ARE NEGLIGIBLE COMPARED TO H^+ IONS FROM ACID.

PLUG IN THE EQUILIBRIUM VALUES IN THE EQUATION FOR K_a

$$K_a = \frac{[CH_3CO_2^-][H^+]}{[CH_3CO_2H]} = \frac{(x)(x)}{(0.8 - x)} = 1.75 \times 10^{-5} \text{ (FROM THE TABLE)}$$

$$\frac{x^2}{0.8} = 1.75 \times 10^{-5}$$

ASSUMPTION 2: x IS NEGLIGIBLY SMALL COMPARED TO 0.8, SO WE CAN IGNORE IT IN THE DENOMINATOR.

$$x^2 = = (0.8)(1.75)10^{-5} = 14 \times 10^{-6}$$

$$x = (14)^{1/2} \times 10^{-3} = 3.74 \times 10^{-3}$$

BUT $x = [H^+]$, SO

$$pH = -\log(3.74 \times 10^{-3}) = 3 - \log(3.74) = 3 - 0.57$$

$$= 2.43$$

ASSUMPTION 2 WAS JUSTIFIED. x REALLY IS MUCH SMALLER THAN 0.8.

THIS ALSO TELLS US THE FRACTION OF MOLECULES THAT IONIZE.

$$\frac{[CH_3CO_2^-]}{[CH_3CO_2H]} = \frac{3.74 \times 10^{-3}}{0.8} = 4.7 \times 10^{-3}$$

A LITTLE LESS THAN 5 MOLECULES IN A THOUSAND.

TRY DOING THE SAME CALCULATION WITH A 0.08 M SOLUTION. MAKE THE SAME TWO SIMPLIFYING ASSUMPTIONS. YOU SHOULD FIND pH = 2.93, AND ALSO THAT THE FRACTION OF IONIZED MOLECULES GOES UP AS CONCENTRATION GOES DOWN.

REACTIONS SUCH AS

$$Fe^{3+} + 2H_2O \rightleftharpoons FeOH^{2+} + H_3O^+$$

ARE CALLED **HYDROLYSIS,** OR WATER-SPLITTING. HERE IT INVOLVES AN ACID, BUT IT'S ALSO VERY COMMON WITH BASES.

WHEN A BASE B^- (OTHER THAN OH^-) IS DISSOLVED IN WATER, B^- TAKES H^+ FROM H_3O^+.

$[H_3O^+]$ DROPS... $[OH^-]$ MUST RISE TO MAINTAIN K_w...

QUICK! SOMEBODY MAKE MORE OH^-!

YOU DO IT!

NO, YOU DO IT!

YOU!

OH, ALL RIGHT...

THIS CAN ONLY HAPPEN BY SPLITTING H_2O, WHICH MAKES MORE H^+...

POP

WHICH IS GOBBLED UP BY B^-... AND SO ON, UNTIL EQUILIBRIUM IS REACHED.

IN OTHER WORDS, B^- **HYDROLYZES** WATER AND CAUSES A RISE IN OH^-.

$$H_2O + B^- \rightleftharpoons HB + OH^-$$

AND WE GET A NEW EQUILIBRIUM CONSTANT, THE **BASE IONIZATION CONSTANT** K_b.

$$K_b = \frac{[HB][OH^-]}{[B^-]}$$

THE HIGHER THE K_b, THE STRONGER THE BASE. THIS IS BECAUSE:

- HIGHER K_b MEANS HIGHER $[OH^-]$, HENCE HIGHER pH.

- K_b MEASURES B^-'S ABILITY TO TAKE A PROTON FROM H_2O.

- K_b IS INVERSE TO K_a. IF HB IS THE CONJUGATE ACID, THEN

BASE B		K_b
OH^-	HYDROXIDE	55.6
S^{2-}	SULFIDE	10^5
CO_3^{2-}	CARBONATE	2.0×10^{-4}
NH_3	AMMONIA	1.8×10^{-5}
$B(OH)_4^-$	BORATE	2.0×10^{-5}
HCO_3^-	BICARBONATE	2.0×10^{-8}

$$K_a K_b = \frac{[H^+][\cancel{B^-}]}{[\cancel{HB}]} \; \frac{[\cancel{HB}][OH^-]}{[\cancel{B^-}]} = [H^+][OH^-] = K_w = 10^{-14}$$

Example.

WHAT'S THE pH OF A 0.15 M SOLUTION OF AMMONIA, NH_3? CALCULATE AS BEFORE, USING THE REACTION

$$NH_3 + H_2O \rightleftharpoons NH_4^+ + OH^-$$

	NH_3	NH_4^+	OH^-
INITIAL CON.	0.15	0.0	0.0
CHANGE IN CON.	$-x$	x	x
EQUILIBRIUM CON.	$0.15-x$	x	x

ASSUMPTION 1: OH^- FROM WATER IS NEGLIGIBLE.

$$K_b = \frac{[NH_4^+][OH^-]}{[NH_3]} \qquad \frac{x^2}{(0.15-x)} = 1.8 \times 10^{-5}$$

ASSUMPTION 2: x IS NEGLIGIBLE COMPARED TO 0.15.

$$\frac{x^2}{0.15} = 1.8 \times 10^{-5}$$

$x^2 = 2.7 \times 10^{-6}$ $\quad x = 1.64 \times 10^{-3}$

$[OH^-] = 1.64 \times 10^{-3}$

$pOH = 3 - \log(1.64) = 2.78$

$pH = 14 - pOH = $ **11.22**

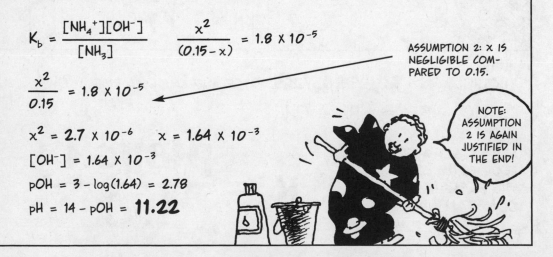

NOTE: ASSUMPTION 2 IS AGAIN JUSTIFIED IN THE END!

Neutralization and Salts

IN WATER, ACIDS GENERATE H^+ AND BASES GENERATE OH^-. WHEN ACIDS AND BASES COMBINE, THESE IONS **NEUTRALIZE** EACH OTHER. FOR EXAMPLE:

$$HCl(aq) + NaOH(aq) \longrightarrow Na^+(aq) + Cl^-(aq) + H_2O$$

TWO NASTY CHEMICALS COMBINE TO MAKE AN ORDINARY SOLUTION OF TABLE SALT IN WATER. IF THE WATER EVAPORATES, ONLY SALT CRYSTALS REMAIN.

THIS IS TYPICAL, SO TYPICAL, IN FACT, THAT IT'S THE DEFINITION OF A SALT: A **SALT** IS A SUBSTANCE FORMED BY THE NEUTRALIZATION OF AN ACID BY A BASE.

BY NEUTRALIZE, WE MEAN THAT THE SALTS ARE MADE FROM EQUIVALENT WEIGHTS OF ACID AND BASE.

AN **EQUIVALENT WEIGHT** OF ACID IS THE AMOUNT THAT WOULD YIELD **ONE MOLE OF PROTONS** IN WATER IF THE ACID IONIZED COMPLETELY.

1 EQUIV HCl = 1 MOL

BUT

1 EQUIV H_2SO_4 = 0.5 MOL

BECAUSE H_2SO_4 CAN GIVE UP TWO PROTONS. SIMILARLY,

1 EQUIV H_2CO_3 = 0.5 MOL

AN EQUIVALENT OF BASE IS THE AMOUNT THAT WOULD GIVE UP ONE MOLE OF OH^- IF THE BASE WERE TO IONIZE COMPLETELY. SO

1 EQUIV $NaOH$ = 1 MOL
1 EQUIV $Ca(OH)_2$ = 0.5 MOL
1 EQUIV NH_3 = 1 MOL

BECAUSE

$$NH_3 + H_2O \longrightarrow NH_4^+ + OH^-$$

IF IT WERE TO IONIZE COMPLETELY.

N EQUIVALENTS OF ACID ALWAYS NEUTRALIZE **N** EQUIVALENTS OF BASE, BECAUSE THEY MAKE EQUAL NUMBERS OF PROTONS AND HYDROXIDE IONS, RESPECTIVELY.

NOTE: A "NEUTRALIZED" SOLUTION MAY NOT BE NEUTRAL! THAT IS, THE pH OF A SALT SOLUTION NEED NOT BE 7.

BUT pH **IS 7** WHENEVER A **STRONG** ACID NEUTRALIZES A **STRONG** BASE, AS WHEN $NaOH$ NEUTRALIZES H_2SO_4 TO MAKE Na_2SO_4. THE SALT IONS HAVE NO ACID OR BASIC EFFECT. THAT'S WHAT IT MEANS THAT THEIR "PARENT" ACID AND BASE WERE STRONG.

WHEN A STRONG ACID NEUTRALIZES A WEAK BASE, THE SOLUTION WILL HAVE pH < 7. CONSIDER AMMONIUM NITRATE, NH_4NO_3, A COMMON INGREDIENT IN FERTILIZER. IT RESULTS FROM THE NEUTRALIZATION OF NH_3 (WEAK BASE) BY HNO_3 (STRONG ACID).

$$HNO_3(aq) + NH_3(aq) \longrightarrow NH_4^+(aq) + NO_3^-(aq)$$

SO BUSY!

NO_3^- HAS NO BASIC EFFECT (BECAUSE HNO_3 IS STRONG), SO WE CAN IGNORE IT. IT'S A "BYSTANDER ION." BUT NH_4^+ IS A WEAK ACID THAT WILL DISSO-CIATE, WITH $K_a = 5.7 \times 10^{-10}$.

$$NH_4^+(aq) \rightleftharpoons NH_3(aq) + H^+(aq)$$

Example

SUPPOSE THE CONCENTRATION OF NH_4NO_3 IS 0.1 M. WHAT IS THE SOLUTION'S pH? WE MAKE THE USUAL TABLE AND COMPUTATION:

	NH_4^+	NH_3	H^+
CONC. BEFORE IONIZATION	0.1	0.0	0.0
CHANGE IN CONC.	$-x$	x	x
EQUILIBRIUM CONC.	$0.1 - x$	x	x

USUAL ASSUMPTION 1: H^+ FROM WATER IS NEGLIGIBLE.

AT EQUILIBRIUM, K_a IS

$$\frac{[H^+][NH_3]}{[NH_4^+]} = 5.7 \times 10^{-10}$$

MAKING THE USUAL TWO ASSUMPTIONS, WE GET

USUAL ASSUMPTION 2: x IS MUCH LESS THAN 0.1 AND CAN BE IGNORED.

$$\frac{x^2}{0.1} = 5.7 \times 10^{-10}$$

$$x^2 = 5.7 \times 10^{-11} = 57 \times 10^{-12}$$
$$x = [H^+] = 7.55 \times 10^{-6}$$
$$pH = 6 - \log(7.55) = 6 - 0.88$$
$$= \mathbf{5.12}$$

YUP! SLIGHTLY ACID!

RAPID GRO

SIMILARLY, WHEN A STRONG BASE NEUTRALIZES A WEAK ACID, THE RESULTING SALT SOLUTION WILL BE WEAKLY BASIC. FOR EXAMPLE, WHEN NaOH NEUTRALIZES CH_3CO_2H, Na^+ IS A "BYSTANDER ION," WHILE ACETATE, $CH_3CO_2^-$, IS A WEAK BASE. WORK OUT FOR YOURSELF THE pH OF A 0.5 M SOLUTION OF $NaCH_3CO_2$. USE K_b OF $CH_3CO_2^- = 5.7 \times 10^{-10}$.

ANS: pH = 9.23

WE CAN SUMMARIZE THE pH OF SALT SOLUTIONS LIKE THIS:

IF SALT RESULTS FROM NEUTRALIZATION OF	pH
STRONG ACID, STRONG BASE	7
STRONG ACID, WEAK BASE	<7
WEAK ACID, STRONG BASE	>7
WEAK ACID, WEAK BASE	<7 IF $K_a > K_b$ 7 IF $K_a = K_b$ >7 IF $K_a < K_b$

Titration

IS THE PROCESS OF NEUTRALIZING AN UNKNOWN SOLUTION BY DRIPPING ("TITRATING") A KNOWN STRONG ACID OR BASE INTO IT.

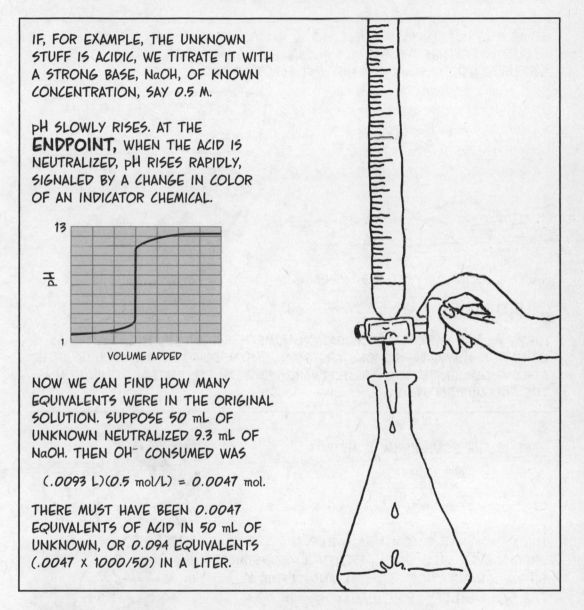

IF, FOR EXAMPLE, THE UNKNOWN STUFF IS ACIDIC, WE TITRATE IT WITH A STRONG BASE, NaOH, OF KNOWN CONCENTRATION, SAY 0.5 M.

pH SLOWLY RISES. AT THE **ENDPOINT,** WHEN THE ACID IS NEUTRALIZED, pH RISES RAPIDLY, SIGNALED BY A CHANGE IN COLOR OF AN INDICATOR CHEMICAL.

NOW WE CAN FIND HOW MANY EQUIVALENTS WERE IN THE ORIGINAL SOLUTION. SUPPOSE 50 mL OF UNKNOWN NEUTRALIZED 9.3 mL OF NaOH. THEN OH⁻ CONSUMED WAS

$$(.0093 \text{ L})(0.5 \text{ mol/L}) = 0.0047 \text{ mol.}$$

THERE MUST HAVE BEEN 0.0047 EQUIVALENTS OF ACID IN 50 mL OF UNKNOWN, OR 0.094 EQUIVALENTS (.0047 x 1000/50) IN A LITER.

CAUTION: THE pH NEED NOT BE 7 AT THE ENDPOINT! THE TITRATION MAY END WITH A SALT THAT HAS ACIDIC OR BASIC PROPERTIES.

WHEN SEVERAL IONS GET TOGETHER IN
SOLUTION, INTERESTING THINGS HAPPEN...

Solubility products

SOME SALTS ARE VERY SOLUBLE, SOME HARDLY AT ALL. WHEN A SALT
SOLUTION REACHES ITS MAXIMUM POSSIBLE CONCENTRATION, WE SAY IT IS
SATURATED. ANY ADDED SALT JUST FALLS TO THE BOTTOM.

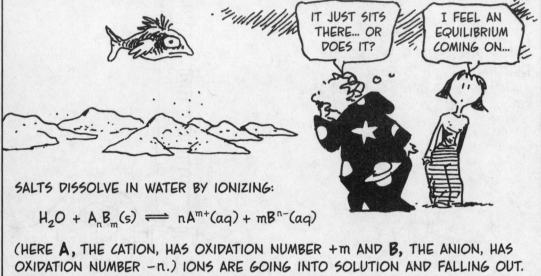

IT JUST SITS
THERE... OR
DOES IT?

I FEEL AN
EQUILIBRIUM
COMING ON...

SALTS DISSOLVE IN WATER BY IONIZING:

$$H_2O + A_nB_m(s) \rightleftharpoons nA^{m+}(aq) + mB^{n-}(aq)$$

(HERE **A**, THE CATION, HAS OXIDATION NUMBER +m AND **B**, THE ANION, HAS
OXIDATION NUMBER −n.) IONS ARE GOING INTO SOLUTION AND FALLING OUT.
AT LOW CONCENTRATION, THE FORWARD REACTION DOMINATES. SATURATION IS
THE EQUILIBRIUM STATE.

HERE IS THE EQUILIBRIUM CONSTANT.

$$K_{eq} = \frac{[A^{m+}]^n[B^{n-}]^m}{[H_2O][A_nB_m]}$$

THE DENOMINATOR CONTAINS WATER AND THE
UNDISSOLVED SALT—BOTH ESSENTIALLY CONSTANT.
SO WE IGNORE THEM AS USUAL AND DEFINE K_{sp},
THE **SOLUBILITY PRODUCT:**

I LOVE
SALT!

$$K_{sp} = [A^{m+}]^n[B^{n-}]^m$$

FOR EXAMPLE, A SATURATED SOLUTION OF $CaCO_3$ HAS A CALCIUM CONCENTRATION OF 6.76×10^{-5} M. POSITIVE AND NEGATIVE CHARGES HAVE TO BALANCE, SO THE CARBONATE CONCENTRATION IS ALSO 6.76×10^{-5} M. THEN:

$$K_{sp} = [Ca^{2+}][CO_3^{2-}]$$
$$= (6.76 \times 10^{-5})^2$$
$$= 4.57 \times 10^{-9}.$$

BECAUSE $CaCO_3$ IS SO INSOLUBLE, WE CAN USE Ca^{2+} IONS TO PRECIPITATE DISSOLVED CO_3^{2-} FROM SOLUTION. FOR INSTANCE, WHEN WE MAKE CAUSTIC LYE, NaOH:

$$Ca(OH)_2(aq) + Na_2CO_3(aq) \longrightarrow 2NaOH + CaCO_3(s)\downarrow$$

Ca^{2+} AND CO_3^{2-} WILL NOT STAY IN SOLUTION TOGETHER BEYOND WHAT THEIR SOLUBILITY PRODUCT ALLOWS. AS SOON AS THE ADDED Ca^{2+} REACHES A LEVEL THAT MAKES

$$[Ca^{2+}][CO_3^{2-}] = 4.57 \times 10^{-9},$$

CALCIUM CARBONATE BEGINS TO PRECIPITATE OUT.

IT DOESN'T TAKE MUCH, IN OTHER WORDS!

EEP!

SOLID	K_{sp}	SOLID	K_{sp}
$FePO_4$	1.26×10^{-18}	$BaSO_4$	10^{-10}
$Fe_3(PO_4)_2$	10^{-33}	$PbCl_2$	1.6×10^{-5}
$Fe(OH)_2$	3.26×10^{-15}	$Pb(OH)_2$	5.0×10^{-15}
FeS	5.0×10^{-18}	$PbSO_4$	1.6×10^{-8}
Fe_2S_3	10^{-88}	PbS	10^{-27}
$Al(OH)_3$ (AMORPH)	10^{-33}	$MgNH_4PO_4$	2.6×10^{-13}
$AlPO_4$	10^{-21}	$MgCO_3$	10^{-5}
$CaCO_3$ (CALCITE)	4.6×10^{-9}	$Mg(OH)_2$	1.82×10^{-11}
$CaCO_3$ (ARAGONITE)	6.0×10^{-9}	$Mn(OH)_2$	1.6×10^{-13}
$CaMg(CO_3)_2$	2.0×10^{-17}	$AgCl$	10^{-10}
CaF_2	5.0×10^{-11}	Ag_2CrO_4	2.6×10^{-12}
$Ca(OH)_2$	5.0×10^{-6}	Ag_2SO_4	1.6×10^{-5}
$Ca_3(PO_4)_2$	10^{-26}	$Zn(OH)_2$	6.3×10^{-18}
$CaSO_4$ (GYPSUM)	2.6×10^{-5}	ZnS	3.26×10^{-22}

K_{sp} CAN HELP US FIND THE EFFECT OF ONE ION ON ANOTHER'S SOLUBILITY. FOR INSTANCE,

pH affects solubility.

Example 1.

$$Ca(OH)_2 \rightleftharpoons Ca^{2+} + 2OH^-$$

$$K_{SP} = [Ca^{2+}][OH^-]^2 = 5.0 \times 10^{-6}$$

TAKE THE LOGARITHM OF BOTH SIDES:

$$\log[Ca^{2+}] + 2\log[OH^-] = (\log 5) - 6$$

$$= 0.7 - 6 = -5.3$$

$$\log[Ca^{2+}] - 2pOH = -5.3$$

SUBSTITUTING $pOH = 14 - pH$,

$$\log[Ca^{2+}] = 22.7 - 2pH$$

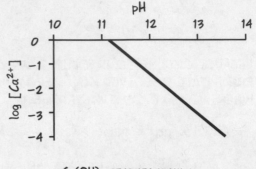

$Ca(OH)_2$ BECOMES HIGHLY SOLUBLE AT pH BELOW 12.

Example 2.

$$CaCO_3 \rightleftharpoons Ca^{2+} + CO_3^{2-}$$

WHEN ACID IS ADDED, CO_3^{2-} TAKES UP H^+ TO MAKE HCO_3^- AND H_2CO_3. HAVING THESE TWO DIFFERENT PRODUCTS COMPLICATES THE MATH, BUT ON BALANCE, THE SITUATION IS DOMINATED BY:

$$H^+ + CO_3^{2-} \rightleftharpoons HCO_3^-$$

BY LE CHATELIER'S PRINCIPLE, ADDING H^+ DRIVES THIS EQUATION TO THE RIGHT AND REMOVES CO_3^{2-}. TO MAINTAIN K_{SP}, MORE $CaCO_3$ WILL DISSOLVE.

TO PRINCIPLE!

BOTH EXAMPLES SHOW HOW LOW-pH WATER TENDS TO DISSOLVE MORE Ca^{2+}. THIS IS A GENERAL PATTERN FOR METALS AND EXPLAINS WHY ACIDIFIED LAKES OFTEN HAVE HIGH LEVELS OF DISSOLVED TOXIC METALS.

Buffers

WE CAN USE BASES' PROTON-CAPTURING PROCLIVITIES TO MODERATE THE pH DROP CAUSED BY STRONG ACIDS.

FOR EXAMPLE, START WITH A LITER OF .01 M SOLUTION OF SODIUM ACETATE, $NaCH_3CO_2$. THIS IONIZES TO GENERATE .01 mol OF THE WEAK BASE ACETATE, $CH_3CO_2^-$, CONJUGATE TO ACETIC ACID.

ADD A LITER OF .01 M HCl, A STRONG ACID. THE ACETATE ION GRABS NEARLY ALL THE PROTONS GIVEN UP BY HCl:

$$CH_3CO_2^- + H^+ \longrightarrow CH_3CO_2H$$

THE pH OF THE SOLUTION IS THAT OF A .005 M SOLUTION OF ACETIC ACID. (CONCENTRATION IS HALVED BECAUSE WE NOW HAVE TWO LITERS OF LIQUID!) THAT'S pH = **3.53.**

OOP!

IF WE HAD ADDED THE HCl TO PURE WATER INSTEAD, THE pH WOULD HAVE DROPPED TO **2.3.** THE ACETATE **MODERATED THE ACIDITY OF THE WATER.**

PROTON-NAPPER!

WE SAY THAT THE ACETATE **BUFFERS** THE SOLUTION AGAINST ACIDS.

IT KEEPS US FROM REALIZING OUR FULL POTENTIAL!

WE MAY BE BOTHERED BY THE FACT THAT OUR BUFFER SOLUTION IS MODERATELY ALKALINE, WITH A pH = 8.38.

WE **ARE** BASES, AFTER ALL!

WE COULD LOWER THIS WITH A WEAK ACID, BUT WE DON'T WANT TO GIVE ANY PROTONS TO THE ACETATE IONS. THIS WOULD CUT THEIR BUFFERING ABILITY.

SO WE **BRILLIANTLY** USE **ACETIC** ACID, CH_3CO_2H. ITS CONJUGATE BASE IS ALREADY ACETATE, SO IT WON'T GIVE UP PROTONS TO THE FREE ACETATE IN SOLUTION.

MY CONJUGATE! WANT A PROTON?

NO MORE THAN YOU DO...

IF WE MAKE A SOLUTION 0.01 M IN ACETATE AND JUST 0.002 M IN ACETIC ACID, THE pH WILL BE 5.5, NOT TOO BAD. (THE CALCULATION IS ON THE PAGE AFTER NEXT.)

EVEN BETTER, WE HAVE BUFFERED AGAINST **ACIDS AND BASES SIMULTANEOUSLY!** THE ACETIC ACID WILL GIVE UP ITS H TO A STRONG BASE, WHILE THE ACETATE WILL TAKE PROTONS FROM STRONG ACIDS. pH WILL BE HELD WITHIN A LIMITED RANGE.

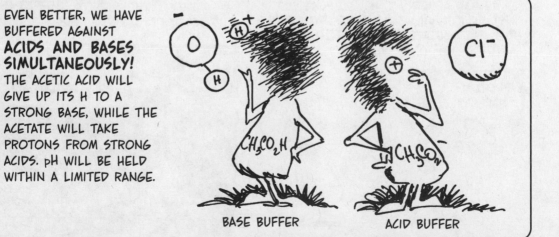

BASE BUFFER   ACID BUFFER

THIS IS THE TRICK WITH BUFFERS: USE AN ACID AND BASE WITH A **COMMON ION:** COMBINE A WEAK ACID **HB** WITH A SALT THAT IONIZES TO GIVE FREE **B⁻**.

I WISH I'D PATENTED THAT IDEA!

A BIT OF ARITHMETIC LETS US PREDICT THE pH OF BUFFERS, BOTH BEFORE AND AFTER ADDITION OF ACIDS OR BASES. WE START WITH THE WEAK ACID HB.

BY DEFINITION,

$$K_a = \frac{[H^+][B^-]}{[HB]}$$

SO

$$\frac{K_a}{[H^+]} = \frac{[B^-]}{[HB]}$$

TAKING LOG OF BOTH SIDES,

$$\log K_a - \log [H^+] = \log ([B^-]/[HB])$$

WRITING pK_a FOR $-\log K_a$, THIS BECOMES

$$\textbf{pH - pK}_a = \textbf{log ([B}^-\textbf{]/[HB])}$$

WHICH IS CALLED THE

Henderson-Hasselbalch Equation.

MAN! TALK ABOUT THE POWER OF H!

IN OUR BUFFER SOLUTION, THE SALT CONCENTRATION GIVES $[B^-]$, AND THE CONCENTRATION OF ACID GIVES $[HB]$. K_a WE KNOW, SO WE CAN SOLVE FOR pH.

FOR EXAMPLE, SUPPOSE A BUFFER SOLUTION CONSISTS OF 1 L OF 0.5 M $NaCH_3CO_2$ AND 0.1 M CH_3CO_2H. K_a OF ACETIC ACID IS 1.75×10^{-5}, SO

$$pK_a = -\log(1.75 \times 10^{-5})$$
$$= 4.76$$

THEN BY HENDERSON-HASSELBALCH, THE pH OF THE SOLUTION IS

$$pH = pK_a + \log([B^-]/[HB])$$
$$= 4.76 + \log(0.5/0.1)$$
$$= 4.76 + \log 5$$
$$= 4.76 + 0.70 = \mathbf{5.46}$$

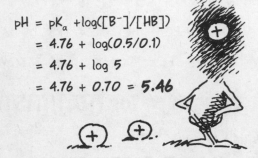

IF A LITER OF 0.05 M HCl IS ADDED, WE ASSUME THAT THE $CH_3CO_2^-$ BINDS WITH ESSENTIALLY ALL THE H^+ FROM HCl:

$$CH_3CO_2^- + H^+ \longrightarrow CH_3CO_2H$$

THEN WE MAKE THE USUAL TABLE:

	CH_3CO_2H	$CH_3CO_2^-$	H^+
ORIG. CON.	0.05	0.25	0.025
CON. CHANGE	0.025	-0.025	-0.025
EQUILIB. CON.	0.075	0.225	0.0

NOTE THAT CONCENTRATIONS ARE HALVED, BECAUSE WE NOW HAVE **TWO LITERS** OF SOLUTION. THEN HENDERSON-HASSELBALCH SAYS:

$$pH = pK_a + \log \frac{[CH_3CO_2^-]}{[CH_3CO_2H]}$$

$$= 4.76 + \log (0.225/0.075)$$
$$= 4.76 + \log 3 = 4.76 + 0.48$$
$$= \mathbf{5.24}$$

AMAZING! IF WE HAD ADDED A LITER OF 0.05 M HCl TO A LITER OF PURE WATER, THE pH WOULD HAVE BEEN LESS THAN **2**!

SEE IF YOU CAN DO THE SAME CALCULATION IF WE HAD ADDED A LITER OF 0.04 M NaOH INSTEAD OF THE HCl.

HENDERSON-HASSELBALCH CAN ALSO GUIDE US WHEN WE WANT TO ADJUST THE pH OF A SYSTEM.

FOR EXAMPLE, NH_4^+ IS MUCH LESS POISONOUS TO FISH THAN NH_3 BECAUSE THE UNCHARGED MOLECULE CAN PASS THROUGH CELL MEMBRANES EASILY AND INTERFERE WITH METABOLISM. HENDERSON-HASSELBALCH SAYS

$$\log ([NH_3]/[NH_4^+]) = pH - pK_a$$

IF, FOR EXAMPLE, WE WANT TO MAKE $[NH_3]/[NH_4^+]$ LESS THAN ONE IN A THOUSAND, I.E., ITS LOG < -3, THEN pH MUST BE LOW ENOUGH THAT

$$pH - pK_a < -3$$

SINCE pK_a OF NH_4^+ IS 9.3, ANY pH < 6.3 WILL DO.

SLIGHTLY TART, BUT NOT **TOO** ACID, PLEASE!

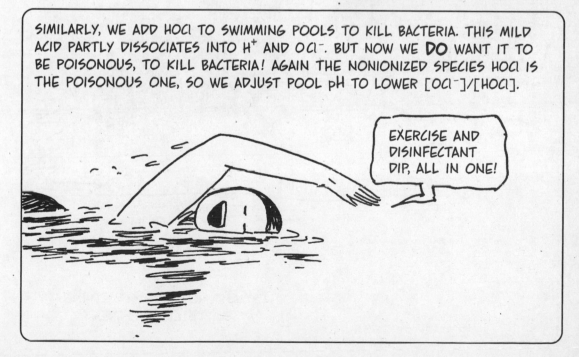

SIMILARLY, WE ADD $HOCl$ TO SWIMMING POOLS TO KILL BACTERIA. THIS MILD ACID PARTLY DISSOCIATES INTO H^+ AND OCl^-. BUT NOW WE **DO** WANT IT TO BE POISONOUS, TO KILL BACTERIA! AGAIN THE NONIONIZED SPECIES $HOCl$ IS THE POISONOUS ONE, SO WE ADJUST POOL pH TO LOWER $[OCl^-]/[HOCl]$.

EXERCISE AND DISINFECTANT DIP, ALL IN ONE!

WE COVERED A LOT IN THIS CHAPTER. WE MET ACIDS AND BASES, MEASURED THEIR STRENGTH, AND SAW HOW THAT STRENGTH IS RELATED TO THEIR IONIZATION IN WATER. WE NEUTRALIZED, TITRATED, AND LOOKED AT THE RESULTING SALTS. WE SAW HOW ACIDS AND BASES AFFECT A SALT'S SOLUBILITY, AND HOW BUFFERS ARE MADE BY COMBINING WEAK ACIDS AND SALTS.

AND NOW FOR SOMETHING COMPLETELY DIFFERENT...

Chapter 10
Chemical Thermodynamics

A HARD, THEORETICAL CHAPTER THAT EXPLAINS
WHY EVERYTHING HAPPENS

WHEN YOU CONTEMPLATE THE UNIVERSE, YOU HAVE TO ADMIT IT LOOKS PRETTY IMPROBABLE. THE SPECTACULAR SPIRALS OF GALAXIES... THE REGAL REGULARITY OF DIAMONDS... THE COMPELLING COMPLEXITY OF LIFE... THE MURKY MYSTERIES OF CHEMISTRY EXPLAINED WITH CARTOONS...

IT'S ALL SO **UNLIKELY!**

THE REASSURING THEME OF THIS CHAPTER IS: THE UNIVERSE GETS **LESS IMPROBABLE** ALL THE TIME.

FOR EXAMPLE, A BRICK FLIES THROUGH A WINDOW, AND THE GLASS SHATTERS AND GOES FLYING.

YOU NEVER SEE A BRICK HIT A PUDDLE OF GLASS FRAGMENTS AND CAUSE THEM TO FLY UP TO MAKE A WINDOW!

OR: SOME AIR IS LET INTO A VACUUM CHAMBER AND QUICKLY FILLS UP THE SPACE.

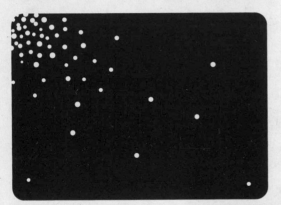

YOU NEVER SEE ALL THE AIR IN A ROOM FLY INTO THE CORNER. (OR IF YOU DO, YOU DON'T LIVE TO TELL THE TALE.)

THE REASON IS THE SAME IN BOTH CASES: THERE ARE MANY, MANY, **MANY** MORE WAYS FOR THINGS TO **FLY APART** OR **SPREAD OUT** THAN THERE ARE FOR THEM TO FLY TOGETHER AND GET CONCENTRATED. SPREADING OUT IS VASTLY MORE PROBABLE. IT'S A GENERAL PRINCIPLE OF THE UNIVERSE:

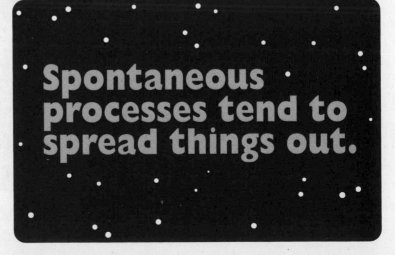

Spontaneous processes tend to spread things out.

YOU MAY OBJECT THAT PICKING UP A BROOM AND SWEEPING THE GLASS SPLINTERS TOGETHER IS A CONCENTRATING PROCESS. AND YOU'D BE RIGHT.

BUT I REPLY THAT IN ORDER TO SWEEP, I HAVE TO MOVE MY BODY. MOVING INVOLVES CHEMICAL REACTIONS THAT SPREAD HEAT INTO THE ENVIRONMENT.

IN FACT, I COULDN'T HAVE MOVED IN THE FIRST PLACE WITHOUT EATING, AND EATING GENERATES WASTE THAT GETS SPREAD AROUND TOO.

THE FOOD I EAT ULTIMATELY DEPENDS ON SOLAR ENERGY, WHICH SPREADS A TERRIFIC AMOUNT OF MATTER AND ENERGY INTO THE UNIVERSE.

YOU HAVE TO LOOK AT THE BIG PICTURE! ANY PROCESS THAT CONCENTRATES MATTER AND/OR ENERGY IN A SYSTEM IS **MORE THAN OFFSET** BY A GREATER AMOUNT OF SPREADING-OUT ELSEWHERE IN THE UNIVERSE. **THE OVERALL EFFECT IN THE UNIVERSE AS A WHOLE IS TO SPREAD THINGS OUT.**

IN CHEMICAL SYSTEMS WE CONSIDER THE SPREADING-OUT OF **ENERGY.**

IMAGINE A SYSTEM CONSISTING OF SOME TYPICALLY HUGE NUMBER OF MOLECULES, AND LET US CONCENTRATE, FOR THE MOMENT, ON ONE OF THEM.

KINETIC ENERGY IS STORED IN A MOLECULE IN THE FORM OF VIBRATION, ROTATION, AND TRANSLATION (I.E., FLYING THROUGH SPACE).

AS WE SAW IN CHAPTER 2, AT THIS SCALE ENERGY IS **QUANTIZED.** ONLY CERTAIN FIXED ENERGY LEVELS ARE ALLOWED.

I OPERATE ON **SO** MANY LEVELS!

ENERGY IS TAKEN ON OR GIVEN OFF IN PACKETS CALLED QUANTA THAT JUMP THE MOLECULE FROM ONE ENERGY LEVEL TO ANOTHER.

ZZAP

SO THIS IS THE PICTURE: EACH MOLECULE HAS ITS OWN ENERGY LEVELS... AND WE THINK OF THE WHOLE SYSTEM AS ALL THESE ENERGY LEVELS TAKEN TOGETHER, WITH A VAST NUMBER OF QUANTA SPREAD OUT AMONG THEM IN SOME WAY.

THESE ADD UP TO THIS!

Entropy, S,

MEASURES THE SPREADING OUT OF ENERGY. IT CAN
BE DEFINED IN TERMS OF HEAT AND TEMPERATURE:

START WITH A SYSTEM AT
TEMPERATURE T (MEASURED
IN °K) AND ADD A SMALL
AMOUNT OF HEAT q.*

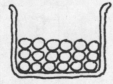

THE **ENTROPY CHANGE**
ΔS, IS GIVEN BY

$$\Delta S = q/T$$

WITH UNITS JOULES/°K.

AS THE FOLLOWING
DIAGRAMS SUGGEST, ΔS
MEASURES THE **EXTRA
SPREADING-OUT OF
HEAT** IN THE SYSTEM
RESULTING FROM THE
ADDITION OF q.

SOMETIMES, q CAUSES A SMALL TEMPERATURE INCREASE
ΔT. (q = CΔT, WHERE C IS THE SYSTEM'S HEAT
CAPACITY.) THE HEAT SPREADS INTO HIGHER ENERGY
LEVELS.

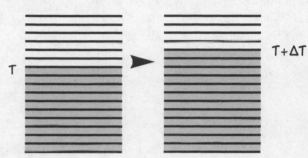

AT OTHER TIMES, q PRODUCES PHASE CHANGE
(MELTING, VAPORIZATION). THEN TEMPERATURE
REMAINS CONSTANT, BUT MOLECULAR MOTION
BECOMES LESS CONSTRAINED, AND MORE LOW-
ENERGY LEVELS "OPEN UP." THE HEAT SPREADS
INTO THESE ENERGY LEVELS.

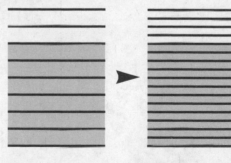

*PHYSICISTS TELL US THAT q MUST BE ADDED **REVERSIBLY,** THAT IS, THE HEAT CAN BE SENT BACK
WITHOUT ANY EXTRA EXPENSE OF ENERGY. THIS IS PHYSICALLY IMPOSSIBLE, BUT CAN BE APPROXIMATELY
ACHIEVED BY ADDING HEAT IN MANY SMALL STEPS.

IT IS NOW POSSIBLE TO
CALCULATE THE **ABSOLUTE
ENTROPY** OF ANY SUBSTANCE.
THIS IS DONE BY ADDING UP
ALL THE LITTLE ENTROPY
INCREMENTS THAT PILE UP AS
THE SUBSTANCE IS HEATED IN
SMALL STEPS FROM ABSOLUTE
ZERO TO SOME CONVENIENT
TEMPERATURE, USUALLY 298°K
(ROOM TEMPERATURE, 25°C).

AT 298°K, WE WRITE S^0, THE
STANDARD ABSOLUTE ENTROPY.

FOR EXAMPLE, FINDING THE STANDARD
ABSOLUTE ENTROPY OF WATER INVOLVES
THESE STEPS:

CHILL A PERFECT ICE CRYSTAL TO
ABSOLUTE ZERO (NOT REALLY POSSIBLE,
BUT CAN BE DONE IN THEORY).

SLOWLY ADD SMALL INCREMENTS OF
HEAT AND ADD UP ALL THE ENTROPY
CHANGES FROM ZERO TO 273°K, THE
MELTING POINT (A TRICKY CALCULATION,
BUT IT CAN BE DONE!). THIS AMOUNTS TO

$$S_{273°} = 47.84 \text{ J/mol°K}$$

MELT THE ICE. WATER'S HEAT OF FUSION
IS 6020 J/MOL, AND T= 273°, SO THE
ADDED ENTROPY HERE IS

$$\frac{6020}{273} = 22.05 \text{ J/mol°K}$$

HEAT LIQUID WATER FROM 273° TO
ROOM TEMPERATURE AND ADD UP THE
ENTROPY CHANGES. THEY TOTAL

$$S_{298°} - S_{273°} = 0.09 \text{ J/mol°K}$$

ADD THE THREE SUBTOTALS FOR THE
**ABSOLUTE STANDARD MOLAR
ENTROPY** OF WATER

$$S^0 (\text{WATER}) = 47.84 + 22.05 + 0.09$$

$$= \textbf{70.0} \text{ JOULES/MOL°K}$$

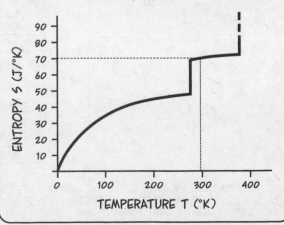

SINCE DIFFERENT SUBSTANCES HAVE DIFFERENT HEAT CAPACITIES AND HEATS OF FUSION AND VAPORIZATION, DIFFERENT AMOUNTS OF HEAT MUST BE ADDED TO RAISE THEIR TEMPERATURES AND CHANGE THEIR STATES. IN OTHER WORDS, EVERY SUBSTANCE HAS ITS OWN CHARACTERISTIC STANDARD ABSOLUTE ENTROPY.

SUBSTANCE	STANDARD MOLAR ENTROPY (J/°K-MOL)
ELEMENTAL SOLIDS	
C (DIAMOND)	2.4
C (GRAPHITE)	5.7
Fe (IRON)	27.3
Cu (COPPER)	33.1
Pb (LEAD)	64.8
IONIC SOLIDS	
CaO	39.7
$CaCO_3$	92.2
$NaCl$	72.3
$MgCl_2$	89.5
$AlCl_3$	167.2
MOLECULAR SOLID	
$C_{12}H_{22}O_{11}$ (SUCROSE)	360.2
LIQUIDS	
H_2O (l)	70
CH_3OH (METHANOL)	126.8
C_2H_5OH (ETHANOL)	161
GASES	
H_2O (g)	189
CH_4 (METHANE)	186
CH_3CH_3 (ETHANE)	230
H_2	131
N_2	191
NH_3	193
O_2	205
CO_2	213
CH_3OH (METHANOL, g)	240
C_2H_5OH (ETHANOL, g)	283

DIAMOND'S AMAZINGLY LOW ENTROPY IS DUE TO ITS HARD, CRYSTALLINE STRUCTURE, WHICH ADMITS VERY LITTLE WIGGLE ROOM. GRAPHITE, MADE OF SHEETS OF ATOMS, HAS MANY MORE ENERGY LEVELS.

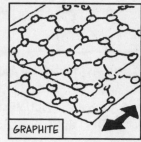

LARGER MOLECULES HAVE HIGHER ENTROPY THAN SMALLER MOLECULES: MORE PARTS TO MOVE.

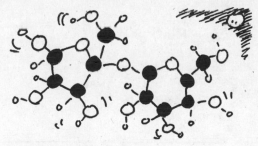

FOR ANY GIVEN SUBSTANCE,

$$S^0(\text{SOLID}) < S^0(\text{LIQUID}) < S^0(\text{GAS}).$$

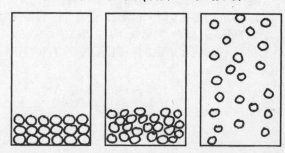

BECAUSE ENTROPY IS RELATED TO SUBSTANCES' COMPOSITION AND INTERNAL STRUCTURE, IT IS POSSIBLE FOR A SYSTEM'S ENTROPY TO CHANGE WITHOUT AN ADDITION OF HEAT. FOR EXAMPLE:

THE **NUMBER** OF PARTICLES IN THE SYSTEM RISES OR FALLS. MORE PARTICLES GENERALLY MEAN MORE ENERGY LEVELS, AND SO ENTROPY GOES UP WITH THE NUMBER OF PARTICLES.

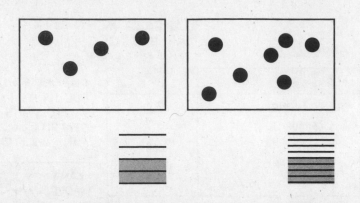

THE SYSTEM **EXPANDS** OR **CONTRACTS**. IT'S A WEIRD QUANTUM-MECHANICAL FACT (TRUST US!) THAT MOLECULES GAIN ENERGY LEVELS WHEN THEY INHABIT A LARGER VOLUME. THEY'RE LIKE DANCERS WHO CAN SHOW OFF MORE MOVES WHEN THERE'S MORE SPACE ON THE FLOOR.

THIS EFFECT EVEN HAS A FORMULA. IF A GAS EXPANDS AT CONSTANT TEMPERATURE, THEN

$$\Delta S = R \ln(P_O/P)$$

WHERE P_O IS THE INITIAL PRESSURE, P IS THE FINAL PRESSURE, AND R IS THE GAS CONSTANT.

THE SYSTEM UNDERGOES A **CHEMICAL REACTION.** A CHEMICAL REACTION CHANGES THE NUMBER OF PARTICLES AND THEIR INTERNAL ARRANGEMENTS. THIS IS SO COMPLICATED IT DESERVES ITS OWN SECTION. SO...

Entropy and Chemical Reactions

THE ENTROPY TABLE IS ONE OF THE CHEMIST'S MOST POWERFUL TOOLS. IT ALLOWS US TO PREDICT WHETHER ANY REACTION WILL GO FORWARD OR NOT (AT STANDARD CONDITIONS).

ENTROPY RULES THE UNIVERSE. WE'VE ALREADY NOTED THAT THE UNIVERSE GOES TOWARDS MORE PROBABLE, SPREAD-OUT STATES. EXPRESSED IN TERMS OF ENTROPY, THIS BECOMES THE FAMOUS **SECOND LAW OF THERMO-DYNAMICS,** WHICH SAYS THAT ENTROPY MUST INCREASE. THAT IS, FOR ANY PROCESS WHATSOEVER,

$$\Delta S_{UNIVERSE} > 0$$

THIS "ONLY" DECIDES EVERYTHING!

FROM THE STANDARD ENTROPY TABLE, WE CAN FIND THE ENTROPY CHANGE OF THE CHEMICALS INVOLVED IN THE REACTION, WHAT WE WILL CALL ΔS_{SYSTEM}:

$$\Delta S_{SYSTEM} = S^0(PRODUCTS) - S^0(REACTANTS)$$

(S IS A "STATE FUNCTION," I.E., IT DEPENDS ONLY ON THE INITIAL AND FINAL STATE OF THE PROCESS AND NOT ON THE STEPS IN BETWEEN.)

AND NOW FOR THE REST OF THE UNIVERSE...

AS AN EXAMPLE, CONSIDER THE HABER PROCESS AT STANDARD CONDITIONS: SUPPOSE WE HAVE A MIXTURE OF N_2, H_2, AND NH_3... THE PARTIAL PRESSURE OF EACH GAS IS 1 ATM, AND T = 298°K. DOES THE REACTION $N_2 + 3H_2 \longrightarrow 2NH_3$ GO FORWARD?

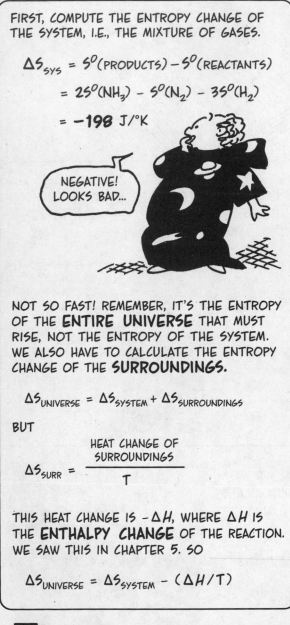

FIRST, COMPUTE THE ENTROPY CHANGE OF THE SYSTEM, I.E., THE MIXTURE OF GASES.

$$\Delta S_{SYS} = S^0 (\text{PRODUCTS}) - S^0 (\text{REACTANTS})$$
$$= 2S^0 (NH_3) - S^0 (N_2) - 3S^0 (H_2)$$
$$= -198 \text{ J/°K}$$

NEGATIVE! LOOKS BAD...

NOT SO FAST! REMEMBER, IT'S THE ENTROPY OF THE **ENTIRE UNIVERSE** THAT MUST RISE, NOT THE ENTROPY OF THE SYSTEM. WE ALSO HAVE TO CALCULATE THE ENTROPY CHANGE OF THE **SURROUNDINGS.**

$$\Delta S_{UNIVERSE} = \Delta S_{SYSTEM} + \Delta S_{SURROUNDINGS}$$

BUT

$$\Delta S_{SURR} = \frac{\text{HEAT CHANGE OF SURROUNDINGS}}{T}$$

THIS HEAT CHANGE IS $-\Delta H$, WHERE ΔH IS THE **ENTHALPY CHANGE** OF THE REACTION. WE SAW THIS IN CHAPTER 5. SO

$$\Delta S_{UNIVERSE} = \Delta S_{SYSTEM} - (\Delta H / T)$$

ΔH FOR THIS REACTION CAN BE READ FROM A TABLE OF ENTHALPIES OF FORMATION. IN FACT, IT'S TWICE ΔH_F OF NH_3 (BECAUSE THERE ARE TWO MOLES PRODUCED):

$$\Delta H = 2\Delta H_F (NH_3)$$
$$= (2 \text{ MOL})(-45.9 \text{ kJ/MOL})$$
$$= -91.8 \text{ kJ}$$

SO

$$\frac{\Delta H}{T} = \frac{-91,800 \text{ J}}{298 \text{ °K}} = -308 \text{ J/°K}$$

THEN THE **TOTAL** ENTROPY CHANGE ASSOCIATED WITH THIS REACTION IS

$$\Delta S_{SYS} - (\Delta H / T)$$
$$= -198 \text{ J/°K} + 308 \text{ J/°K}$$
$$= 110 \text{ J/°K}$$

IT IS POSITIVE! ALTHOUGH THE **SYSTEM'S** ENTROPY FALLS, ENOUGH ENERGY IS SPREAD IN THE **SUR-ROUNDINGS** TO ALLOW THE REACTION TO GO FORWARD!

IT'S ANALOGOUS TO SWEEPING UP BROKEN GLASS. THE PROCESS CONCENTRATES ENERGY WITHIN THE SYSTEM, BUT THE REST OF THE UNIVERSE HAS TO SPREAD OUT ENERGY TO ENABLE IT TO HAPPEN.

THE SAME APPROACH APPLIES TO **ANY** REACTION AT CONSTANT P AND T. IF ΔH IS THE REACTION'S ENTHALPY, THEN

$$\Delta S_{SURROUNDINGS} = -\Delta H/T.$$

THE TOTAL ENTROPY IS

$$\Delta S_{UNIVERSE} = \Delta S_{SYSTEM} + \Delta S_{SURROUNDINGS}$$

WHICH BECOMES

$$\Delta S_{UNIVERSE} = \Delta S_{SYSTEM} - (\Delta H/T)$$

THIS IS THE **TOTAL SPREADING OF ENERGY** IN THE UNIVERSE AS A RESULT OF THE REACTION.

> YOU MIGHT CALL IT THE SYSTEM'S ENTROPY FIGHTING WITH THE ENTHALPY!

BY THE DEFINITION OF ENTROPY, THE TOTAL AMOUNT OF **ENERGY** SPREAD IS $T\Delta S_{UNIVERSE}$. WE SAY THE REACTION HAS A **FREE ENERGY CHANGE** OF $-T\Delta S_{UNIVERSE}$. THIS LAST EXPRESSION IS CALLED ΔG, AFTER THE AMERICAN CHEMIST J. WILLARD GIBBS (1839–1903). MULTIPLYING THE LAST EQUATION BY $-T$ GIVES THIS VALUABLE EXPRESSION FOR ΔG:

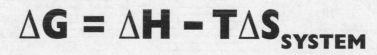

$$\Delta G = \Delta H - T\Delta S_{SYSTEM}$$

A REACTION IS **SPONTANEOUS** WHEN $\Delta S > 0$, IN OTHER WORDS, WHEN $\Delta G < 0$. EQUILIBRIUM COMES WHEN $\Delta G = 0$.

NOTE THAT ΔG IS DESCRIBED STRICTLY IN TERMS OF THE SYSTEM, NOT THE SURROUNDINGS.

WOW!

VERY GOOD!

GIBBS

ΔG REPRESENTS THE NET AMOUNT OF ENERGY THAT CAN POTENTIALLY BE CAPTURED AS WORK WHEN IT SPREADS OUT. IN FACT, YOU CAN THINK OF THE GIBBS FUNCTION AS THE **MAXIMUM AMOUNT OF WORK** THAT CAN BE DONE BY THE REACTION.

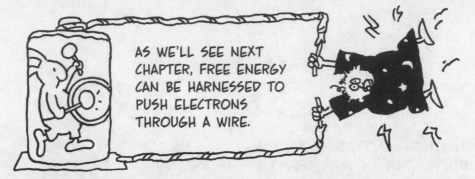

AS WE'LL SEE NEXT CHAPTER, FREE ENERGY CAN BE HARNESSED TO PUSH ELECTRONS THROUGH A WIRE.

YOU CAN THINK OF THE TWO TERMS IN THE GIBBS FUNCTION GRAPHICALLY:

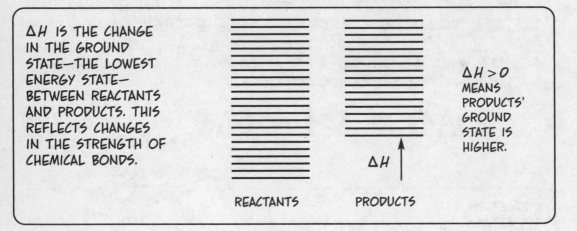

ΔH IS THE CHANGE IN THE GROUND STATE—THE LOWEST ENERGY STATE— BETWEEN REACTANTS AND PRODUCTS. THIS REFLECTS CHANGES IN THE STRENGTH OF CHEMICAL BONDS.

ΔH

$\Delta H > 0$ MEANS PRODUCTS' GROUND STATE IS HIGHER.

REACTANTS   PRODUCTS

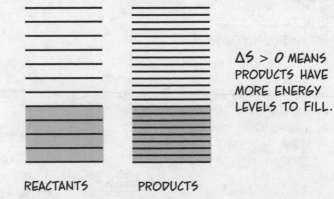

$-T\Delta S$, THE ENERGY ASSOCIATED WITH THE SYSTEM'S ENTROPY CHANGE, REFLECTS CHANGES OF K.E. STATES BETWEEN REACTANTS AND PRODUCTS, I.E., DIFFERENCES OF SIZE, SHAPE, ARRANGEMENT OF MOLECULES, ETC.

$\Delta S > 0$ MEANS PRODUCTS HAVE MORE ENERGY LEVELS TO FILL.

REACTANTS   PRODUCTS

WHEN IS A REACTION SPONTANEOUS? IT HELPS TO DISTINGUISH AMONG FOUR CASES, DEPENDING ON THE SIGNS OF ΔH AND ΔS (MEANING ΔS_{SYSTEM}).

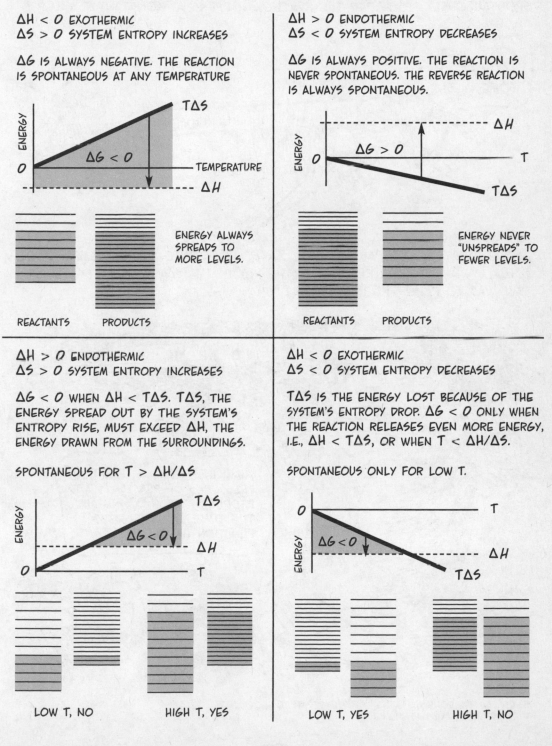

$\Delta H < 0$ EXOTHERMIC
$\Delta S > 0$ SYSTEM ENTROPY INCREASES

ΔG IS ALWAYS NEGATIVE. THE REACTION IS SPONTANEOUS AT ANY TEMPERATURE

ENERGY ALWAYS SPREADS TO MORE LEVELS.

$\Delta H > 0$ ENDOTHERMIC
$\Delta S < 0$ SYSTEM ENTROPY DECREASES

ΔG IS ALWAYS POSITIVE. THE REACTION IS NEVER SPONTANEOUS. THE REVERSE REACTION IS ALWAYS SPONTANEOUS.

ENERGY NEVER "UNSPREADS" TO FEWER LEVELS.

$\Delta H > 0$ ENDOTHERMIC
$\Delta S > 0$ SYSTEM ENTROPY INCREASES

$\Delta G < 0$ WHEN $\Delta H < T\Delta S$. $T\Delta S$, THE ENERGY SPREAD OUT BY THE SYSTEM'S ENTROPY RISE, MUST EXCEED ΔH, THE ENERGY DRAWN FROM THE SURROUNDINGS.

SPONTANEOUS FOR $T > \Delta H/\Delta S$

$\Delta H < 0$ EXOTHERMIC
$\Delta S < 0$ SYSTEM ENTROPY DECREASES

$T\Delta S$ IS THE ENERGY LOST BECAUSE OF THE SYSTEM'S ENTROPY DROP. $\Delta G < 0$ ONLY WHEN THE REACTION RELEASES EVEN MORE ENERGY, I.E., $\Delta H < T\Delta S$, OR WHEN $T < \Delta H/\Delta S$.

SPONTANEOUS ONLY FOR LOW T.

IN OTHER WORDS, THE COMPONENTS OF THE GIBBS FUNCTION, ΔH AND TΔS, PREDICT THE TEMPERATURE RANGE WITHIN WHICH A REACTION WILL TAKE PLACE SPONTANEOUSLY—PROVIDED THE REACTION HAPPENS AT CONSTANT T AND P.

A REASONABLE ASSUMPTION— SOMETIMES!

IN THE HABER PROCESS, AS WE SAW, ΔS < 0, ΔH < 0, SO RAISING TEMPERATURE ACTUALLY INHIBITS THE REACTION.* (THE KEY IN THAT CASE, AS LE CHATELIER SAW, WAS TO RAISE THE PRESSURE.)

YEAH! IN REACTIONS, **PARTIAL** PRESSURES OF REACTANTS AND PRODUCTS ARE CHANGING ALL THE TIME! CAN THE GIBBS FUNCTION HANDLE **THAT?**

*EVEN SO, IT'S DONE AT FAIRLY HIGH TEMPERATURE BECAUSE OF THE FASTER KINETICS AT HIGH T.

TO APPLY GIBBS FREE ENERGY, WE BEGIN WITH A REACTION AT STANDARD CONDITIONS, AND THEN TWEAK THE GIBBS FUNCTION TO REFLECT CHANGES IN PARTIAL PRESSURES OR CONCENTRATIONS.

EVERY SUBSTANCE HAS A **STANDARD FREE ENERGY OF FORMATION** G_F^0. THIS IS THE FREE ENERGY CHANGE WHEN THE SUBSTANCE IS MADE FROM ITS CONSTITUENT ELEMENTS AT STANDARD CONDITIONS. IN OTHER WORDS, IT IS ΔG OF

ELEMENTS $\longrightarrow$ SUBSTANCE

NATURALLY, CHEMISTS HAVE COMPILED VAST TABLES OF THESE. HERE IS A LITTLE ONE.

SUBSTANCE	G_F^0 (kJ/MOL)
CO_2 (g)	−394.37
NH_3 (g)	−16.4
N_2 (g)	0
H_2 (g)	0
CaO (s)	−604.2
H_2O (l)	−237.18
H_2O (g)	−228.59
O_2 (g)	0
H^+ (aq)	0
OH^- (aq)	−157.29

ONE CAN SHOW (AS WITH ENTHALPY OF FORMATION*) THAT **ANY REACTION TAKING PLACE AT STANDARD CONDITIONS** HAS FREE ENERGY EQUAL TO THE DIFFERENCE BETWEEN THE STANDARD FREE ENERGY OF FORMATION OF THE PRODUCTS AND THE STANDARD FREE ENERGY OF FORMATION OF THE REACTANTS:

$$\Delta G = G_F^0(\text{PRODUCTS}) - G_F^0(\text{REACTANTS})$$

CHEMISTS LOVE VAST TABLES!

ACS Annual Free Buffet

*SEE P. 101

LET'S WRITE ΔG^0 TO INDICATE THAT OUR REACTION TAKES PLACE AT STANDARD CONDITIONS (T = 298°K, P = 1 ATM). WHAT HAPPENS WHEN WE CHANGE PRESSURE?

WHEN A GAS CHANGES PRESSURE AT CONSTANT T FROM AN INITIAL PRESSURE P_0 TO A FINAL PRESSURE P, THE ENTROPY CHANGE OBEYS THIS EQUATION (OFFERED WITHOUT PROOF—SORRY!):

$$\Delta S = R \ln(P_0/P) \qquad \text{(R THE GAS CONSTANT)}$$

REMEMBER, EXPANSION INCREASES ENTROPY!

THE PRESSURE CHANGE INVOLVES NO HEAT TRANSFER: $\Delta H = 0$. SO THIS PROCESS (I.E., THE PRESSURE CHANGE) HAS FREE ENERGY:

$$G_F - G_F^0 = \Delta H - T\Delta S = -T\Delta S = -RT\ln(P_0/P)$$

SO

$$G_F = G_F^0 - RT\ln(P_0/P) = G_F^0 + RT\ln(P/P_0)$$

$$= G_F^0 + RT\ln P$$

(BECAUSE $P_0 = 1$ AT STANDARD CONDITIONS).

EXCELLENT! NOW LET P VARY AND CONSIDER REACTIONS AT CONSTANT T = 298°K. THEN

$$\Delta G = G_F(\text{PRODUCTS}) - G_F(\text{REACTANTS})$$

NOW LOOK AT ANY HYPOTHETICAL REACTION WITH BALANCED EQUATION

$$aA + bB \rightleftharpoons cC + dD$$

DOES ANYTHING LOOK FAMILIAR?

AND ASSUME **A**, **B**, **C**, AND **D** ARE ALL GASES THAT REMAIN MIXED TOGETHER, WITH PARTIAL PRESSURES P_A, P_B, P_C, AND P_D. THEN

$$\Delta G = G_F(\text{PRODUCTS}) - G_F(\text{REACTANTS})$$

$$= G_F^0(\text{PROD}) - G_F^0(\text{REAC}) + RT(c\ln P_C + d\ln P_D - a\ln P_A - b\ln P_B)$$

$$= \Delta G^0 + RT\ln\left(\frac{P_C^c P_D^d}{P_A^a P_B^b}\right)$$

Equilibrium Again

$$Q = \frac{P_C^{\,c} P_D^{\,d}}{P_A^{\,a} P_B^{\,b}}$$

IS CALLED THE **REACTION QUOTIENT.** Q IS SMALL WHEN PRODUCTS ARE SCARCE COMPARED TO REACTANTS, AND LARGE WHEN VICE VERSA. IF **A, B, C,** AND **D** ARE DISSOLVED CHEMICALS, WE CAN ALSO WRITE

$$Q = \frac{[C]^c [D]^d}{[A]^a [B]^b}$$

AND IT REMAINS TRUE THAT

$$\Delta G = \Delta G^0 + RT\ln Q$$

NOTE THAT $\Delta G < 0$ IF Q IS SMALL ENOUGH, AND $\Delta G > 0$ IF Q IS LARGE ENOUGH, THAT IS, IF LOTS OF **C** AND **D** ARE PRESENT.

TRANSLATION: WHEN Q IS SMALL, THE REACTION GOES FORWARD! WHEN Q IS LARGE, THE REACTION REVERSES!

THANK YOU.

EQUILIBRIUM OCCURS WHEN $\Delta G = 0$, OR

$$RT\ln Q = -\Delta G^0$$

OR

$$Q = e^{(-\Delta G^0/RT)}$$

K!

EVERYTHING HERE IS CONSTANT!

THIS IS A SECOND DERIVATION OF THE EQUILIBRIUM CONSTANT! IT SAYS THAT AT EQUILIBRIUM, THERE IS A CONSTANT K_{eq} SUCH THAT

$$\frac{[C]^c [D]^d}{[A]^a [B]^b} = K_{eq}$$

AND SIMILARLY FOR PARTIAL PRESSURES. EVEN BETTER, NOW WE CAN CALCULATE K_{eq} FROM STANDARD FREE ENERGIES OF FORMATION, WITHOUT EVER RUNNING THE REACTION!

$$K_{eq} = e^{(-\Delta G^0/RT)}$$

(AND REMEMBER, IN THIS EQUATION $T = 298°K$.)

JUST FOR FUN, LET'S SEE IF WE CAN
CALCULATE THE IONIZATION CONSTANT OF
WATER IN THIS WAY.

$$H_2O \; (l) \rightleftharpoons H^+(aq) + OH^-(aq)$$

$$\Delta G^0 = G^0_F (\text{PRODUCTS}) - G^0_F (\text{REACTANTS})$$

FROM THE TABLE:

$$G^0_F (H_2O \; (l)) = -237.18 \; kJ/mol$$

$$G^0_F (OH^-(aq)) = -157.29 \; kJ/mol$$

$$G^0_F (H^+(aq)) = 0$$

SO

$$\Delta G^0 = -157.29 - (-237.18) = 79.89 \; kJ/mol$$

$$= 79,890 \; J/mol$$

$$K_{eq} = e^{(-\Delta G^0/RT)}$$

$$= e^{(-79,890)/(8.3134)(298)}$$

$$= e^{-32.25}$$

$$= 9.9 \times 10^{-15}$$

$$= 10^{-14} \quad \text{OR CLOSE ENOUGH!}$$

AMAZING! IT WORKED!

OF COURSE IT WORKED...

NO, IT'S SHOCKING, REALLY!

IF YOU THINK **THAT'S** SHOCKING, WAIT UNTIL YOU SEE **THIS**...

Chapter 11
Electrochemistry

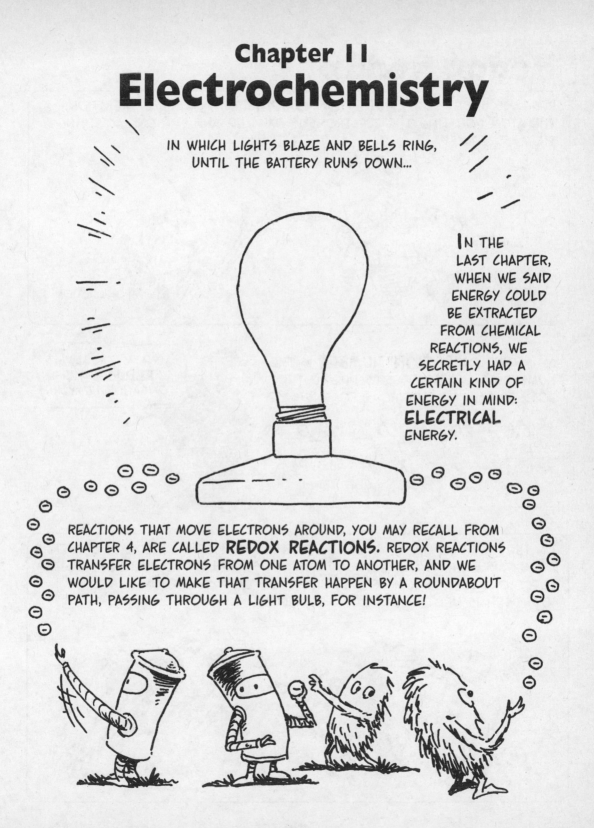

IN WHICH LIGHTS BLAZE AND BELLS RING,
UNTIL THE BATTERY RUNS DOWN...

IN THE LAST CHAPTER, WHEN WE SAID ENERGY COULD BE EXTRACTED FROM CHEMICAL REACTIONS, WE SECRETLY HAD A CERTAIN KIND OF ENERGY IN MIND: **ELECTRICAL** ENERGY.

REACTIONS THAT MOVE ELECTRONS AROUND, YOU MAY RECALL FROM CHAPTER 4, ARE CALLED **REDOX REACTIONS.** REDOX REACTIONS TRANSFER ELECTRONS FROM ONE ATOM TO ANOTHER, AND WE WOULD LIKE TO MAKE THAT TRANSFER HAPPEN BY A ROUNDABOUT PATH, PASSING THROUGH A LIGHT BULB, FOR INSTANCE!

Redox Redux

REDOX IS SHORT FOR **REDUCTION-OXIDATION.** IN A REDOX REACTION, THE ATOM DONATING THE ELECTRONS IS OXIDIZED, AND THE ONE ACCEPTING THEM IS REDUCED.

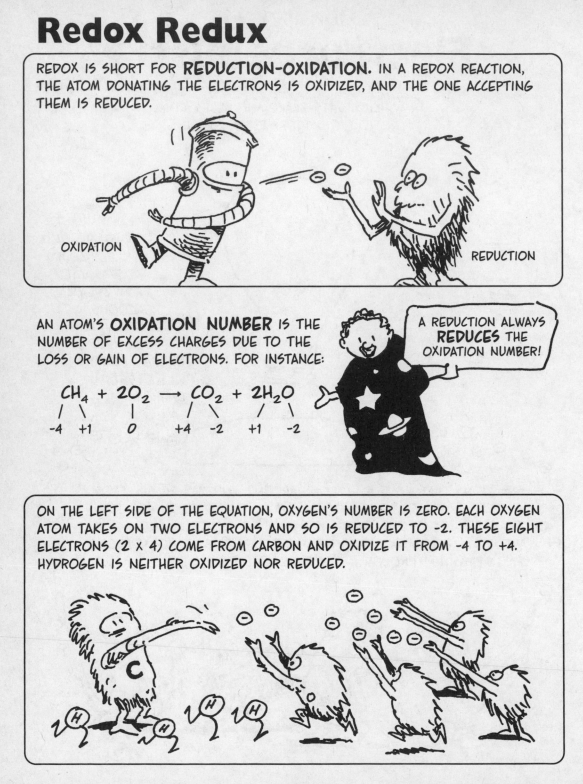

OXIDATION

REDUCTION

AN ATOM'S **OXIDATION NUMBER** IS THE NUMBER OF EXCESS CHARGES DUE TO THE LOSS OR GAIN OF ELECTRONS. FOR INSTANCE:

$$CH_4 + 2O_2 \longrightarrow CO_2 + 2H_2O$$

-4 +1   0   +4 -2   +1 -2

A REDUCTION ALWAYS **REDUCES** THE OXIDATION NUMBER!

ON THE LEFT SIDE OF THE EQUATION, OXYGEN'S NUMBER IS ZERO. EACH OXYGEN ATOM TAKES ON TWO ELECTRONS AND SO IS REDUCED TO -2. THESE EIGHT ELECTRONS (2 x 4) COME FROM CARBON AND OXIDIZE IT FROM -4 TO +4. HYDROGEN IS NEITHER OXIDIZED NOR REDUCED.

IN CHAPTER 4, WE SAW OXIDATIONS PERFORMED MOSTLY BY NON-METALS LIKE OXYGEN, BUT REDOX REACTIONS ARE ALSO COMMON AMONG **METALS AND THEIR IONS.** FOR EXAMPLE, ZINC SHEDS ELECTRONS MORE READILY THAN COPPER. WHEN Zn MEETS A Cu^{2+} ION, TWO ELECTRONS JUMP FROM ZINC TO COPPER. Cu^{2+} **OXIDIZES** Zn, AND Zn **REDUCES** Cu^{2+}.

$$Zn + Cu^{2+} \longrightarrow Zn^{2+} + Cu$$

IF A ZINC BAR IS IMMERSED IN A SOLUTION OF COPPER (II) SULFATE,* $CuSO_4$, THE ZINC METAL SLOWLY OXIDIZES AND DISSOLVES, WHILE COPPER IONS PICK UP ELECTRONS AND FALL OUT OF SOLUTION AS PURE METALLIC COPPER.

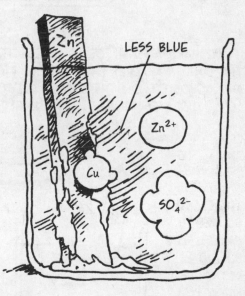

 IN THIS REACTION, ELECTRONS MOVE STRAIGHT FROM ONE ATOM OR ION TO ANOTHER. BUT NOW WE DO SOMETHING CLEVER: **SEPARATE** THE OXIDATION FROM THE REDUCTION, BUT CONNECT THE REACTION SITES BY A CONDUCTING WIRE.

*IT'S BLUE, BY THE WAY!

A ZINC BAR IS IMMERSED IN A 1M AQUEOUS SOLUTION OF $ZnSO_4$. COPPER IS IMMERSED IN A 1M SOLUTION OF $CuSO_4$. THE TWO BARS—OR **ELECTRODES**— ARE CONNECTED BY A WIRE. ELECTRONS WILL STILL NOT FLOW, HOWEVER, SINCE THEY WOULD CREATE A CHARGE IMBALANCE.

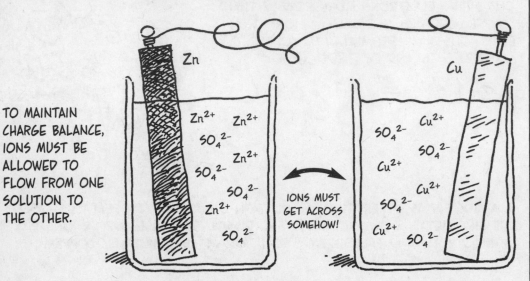

TO MAINTAIN CHARGE BALANCE, IONS MUST BE ALLOWED TO FLOW FROM ONE SOLUTION TO THE OTHER.

IONS MUST GET ACROSS SOMEHOW!

IF WE MAKE A PATH FOR IONS, ELECTRONS WILL MOVE THROUGH THE WIRE. IT'S THE ONLY WAY THEY CAN GET FROM Zn TO Cu^{2+}! DISSOLVED Cu^{2+} IS REDUCED AND DEPOSITED ON THE COPPER ELECTRODE. Zn IS OXIDIZED AND DISSOLVES. SO_4^{2-} MIGRATES TOWARD THE ZINC ELECTRODE. $[Zn^{2+}]$ RISES AND $[Cu^{2+}]$ FALLS.

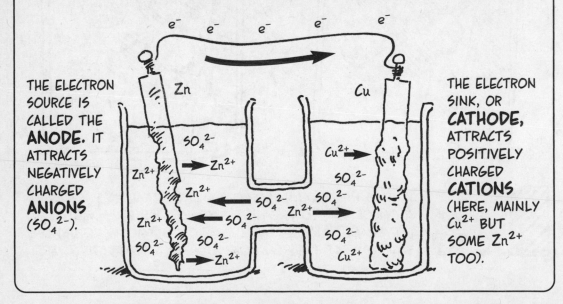

THE ELECTRON SOURCE IS CALLED THE **ANODE.** IT ATTRACTS NEGATIVELY CHARGED **ANIONS** (SO_4^{2-}).

THE ELECTRON SINK, OR **CATHODE,** ATTRACTS POSITIVELY CHARGED **CATIONS** (HERE, MAINLY Cu^{2+} BUT SOME Zn^{2+} TOO).

WHY DO THE ELECTRONS FLOW? BECAUSE FOR THEM IT'S LIKE FALLING DOWNHILL! THE ELECTRONS HAVE A **LOWER POTENTIAL ENERGY** AT THE CATHODE. TO PUT IT ANOTHER WAY, ENERGY WOULD HAVE TO BE ADDED FROM OUTSIDE TO PUSH THE ELECTRONS "UPHILL" FROM CATHODE TO ANODE.

NOTE: THIS IS AN **ANALOGY** ONLY! ELECTRONS ARE **NOT** LITERALLY FLOWING DOWNHILL! JUST **LOSING ENERGY!**

THE REACTION'S "PUSH"—THE ENERGY DROP PER CHARGE—IS CALLED THE **VOLTAGE** OR **ELECTRIC POTENTIAL**, ΔE. ITS UNITS ARE **VOLTS**, ABOUT WHICH MORE SOON. A METER ON THE WIRE SHOWS THAT THE COPPER-ZINC REACTION GENERATES **1.1 VOLTS**. WE CAN HARNESS THIS "ELECTRON SPILLWAY" WITH A LIGHT BULB OR MOTOR OR BELL. THE ELECTRONS **DO WORK.**

EUREKA! EUREKA! EUREKA!

THIS SETUP IS CALLED A **VOLTAIC CELL,** OR LOOSELY SPEAKING, AN ELECTRIC **BATTERY.***

*STRICTLY SPEAKING, A BATTERY CONSISTS OF SEVERAL CELLS WIRED IN SERIES.

BECAUSE A CHEMICAL CELL PHYSICALLY SEPARATES REDUCTION AND OXIDATION, CHEMISTS LIKE TO THINK IN TERMS OF SEPARATE **HALF-REACTIONS** THAT DESCRIBE THE ELECTRON TRANSFERS. IN THE ZINC-COPPER CELL, THE HALF-REACTIONS ARE:

OXIDATION: $Zn \longrightarrow Zn^{2+} + 2e^-$
REDUCTION: $Cu^{2+} + 2e^- \longrightarrow Cu$

WHEN HALF-REACTIONS ARE ADDED TOGETHER, ELECTRONS APPEAR ON BOTH SIDES AND CAN BE CANCELLED:

$Zn + Cu^{2+} + \cancel{2e^-} \longrightarrow Zn^{2+} + Cu + \cancel{2e^-}$

MORE (SIMPLE) REDOX REACTIONS IN SOLUTION AND THEIR HALF REACTIONS:

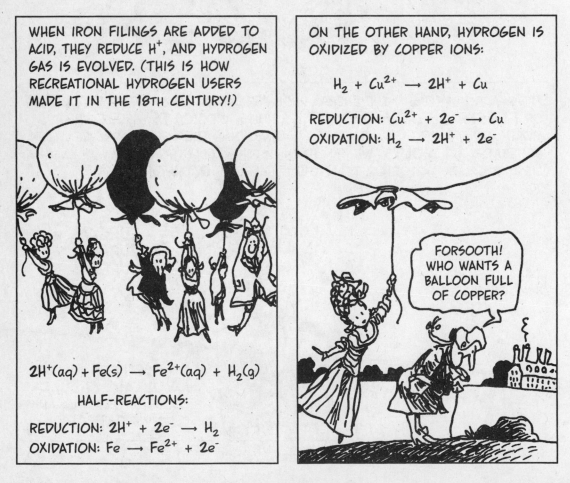

WHEN IRON FILINGS ARE ADDED TO ACID, THEY REDUCE H^+, AND HYDROGEN GAS IS EVOLVED. (THIS IS HOW RECREATIONAL HYDROGEN USERS MADE IT IN THE 18TH CENTURY!)

$2H^+(aq) + Fe(s) \longrightarrow Fe^{2+}(aq) + H_2(g)$

HALF-REACTIONS:

REDUCTION: $2H^+ + 2e^- \longrightarrow H_2$
OXIDATION: $Fe \longrightarrow Fe^{2+} + 2e^-$

ON THE OTHER HAND, HYDROGEN IS OXIDIZED BY COPPER IONS:

$H_2 + Cu^{2+} \longrightarrow 2H^+ + Cu$

REDUCTION: $Cu^{2+} + 2e^- \longrightarrow Cu$
OXIDATION: $H_2 \longrightarrow 2H^+ + 2e^-$

FORSOOTH! WHO WANTS A BALLOON FULL OF COPPER?

LISTING ΔE FOR EVERY REDOX REACTION WOULD BE TEDIOUS, BUT IT TURNS OUT WE CAN ASSIGN VOLTAGES E_{OX} AND E_{RED} TO THE HALF-REACTIONS AND ADD THEM TOGETHER.

$$\Delta E = E_{OX} + E_{RED}$$

THE VOLTAGE OF ANY FULL REACTION IS FOUND BY ADDING UP ITS HALF-REACTION POTENTIALS. MUCH MORE CONVENIENT!

NO CHEMIST IS IMMUNE TO THE BEAUTY OF AN IMPROVED BOOKKEEPING SCHEME...

TABLES

MORE TABLES

SO, FOR INSTANCE,

$$E_{OX}(Zn \longrightarrow Zn^{2+} + 2e^-) = 0.76V$$
$$E_{RED}(Cu^{2+} + 2e^- \longrightarrow Cu) = 0.34V$$

ΔE OF THE WHOLE REACTION IS

$$0.77 + 0.34 = 1.10V$$

BUT WHERE DID THESE NUMBERS COME FROM, ANYWAY?

WE CAN THINK OF THESE AS THE OXIDIZED SPECIES' TENDENCY TO GIVE ELECTRONS AWAY AND THE REDUCED SPECIES' URGE TO PICK THEM UP.

GIVE AWAY TO WHOM? PICK UP FROM WHERE?

HOW CAN WE ASSIGN VOLTAGES TO HALF-REACTIONS WHEN HALF-REACTIONS NEVER HAPPEN ALONE?

THIS IS HOW: FIRST, SINCE VOLTAGE DEPENDS ON CONCENTRATION, PRESSURE, AND TEMPERATURE, WE ASSUME **STANDARD CONDITIONS:** T = 298°K, P = 1 ATM, CONCENTRATION = 1 M. WE CALL OUR HALF-REACTION VOLTAGE A **STANDARD REDUCTION POTENTIAL,** E^o_{RED}, OR SIMPLY E^o.

IS THERE ANYTHING THAT **DOESN'T** DEPEND ON TEMPERATURE, PRESSURE, AND CONCENTRATION?

IT WILL BE A REDUCTION POTENTIAL, BECAUSE FOR CONVENIENCE **WE WRITE ALL HALF-REACTIONS AS REDUCTIONS.** IF A REACTION RUNS LEFT TO RIGHT, IT'S A REDUCTION; IF RIGHT TO LEFT, IT'S AN OXIDATION, AND

$$E_{RED} = -E_{OX}.$$

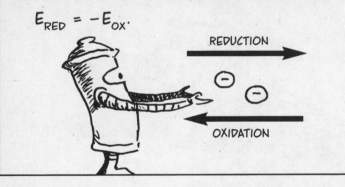

REDUCTION

OXIDATION

FINALLY, WE MEASURE ALL REDUCTION POTENTIALS AGAINST THAT OF **HYDROGEN,** I.E., THE REDUCTION $2H^+ + 2e^- \longrightarrow H_2$, WHICH IS ASSIGNED A VALUE $E^o = 0$.

THE HYDROGEN REDUCTION IS DONE BY BUBBLING H_2 AT ONE ATM OVER A CATALYST, PLATINUM DIOXIDE, PtO_2, INTO AN ACID AT pH=0 (AT STANDARD CONDITIONS, $[H^+]$ = 1 M).

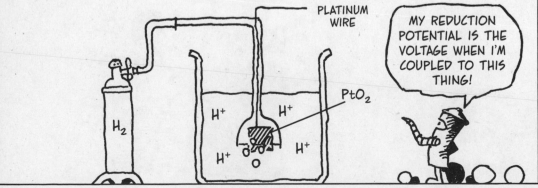

PLATINUM WIRE

H_2

H^+ H^+

H^+ H^+

PtO_2

MY REDUCTION POTENTIAL IS THE VOLTAGE WHEN I'M COUPLED TO THIS THING!

SOME HALF-REACTIONS OXIDIZE H_2 (E.G., $Cu^{2+} + 2e^- \longrightarrow Cu$), WHILE OTHERS ($Fe^{2+} + 2e^- \longrightarrow Fe$) REDUCE H^+. ANYTHING THAT REDUCES H^+ WILL HAVE A **NEGATIVE REDUCTION POTENTIAL.**

HALF-REACTION	E^0 (V)	HALF-REACTION	E^0 (V)
$Li^+ + e^- \longrightarrow Li$	-3.05	$Ni^{2+} + 2e^- \longrightarrow Ni$	-0.25
$K^+ + e^- \longrightarrow K$	-2.93	$Sn^{2+} + 2e^- \longrightarrow Sn$	-0.14
$Ba^{2+} + 2e^- \longrightarrow Ba$	-2.92	$Pb^{2+} + 2e^- \longrightarrow Pb$	-0.13
$Sr^{2+} + 2e^- \longrightarrow Sr$	-2.89	$2H^+ + 2e^- \longrightarrow H_2$	0.00
$Ca^{2+} + 2e^- \longrightarrow Ca$	-2.84	$AgCl(s) + e^- \longrightarrow Ag(s) + Cl^-$	0.22
$Na^+ + e^- \longrightarrow Na$	-2.71	$Cu^{2+} + 2e^- \longrightarrow Cu$	0.34
$Mg^{2+} + 2e^- \longrightarrow Mg$	-2.38	$O_2 + 2H_2O + 4e^- \longrightarrow 4OH^-$	0.40
$Be^{2+} + 2e^- \longrightarrow Be$	-1.85	$Cu^+ + e^- \longrightarrow Cu$	0.52
$Al^{3+} + 3e^- \longrightarrow Al$	-1.66	$I_2 + 2e^- \longrightarrow 2I^-$	0.54
$Ti^{2+} + 2e^- \longrightarrow Ti$	-1.63	$Fe^{3+} + e^- \longrightarrow Fe^{2+}$	0.77
$Mn^{2+} + 2e^- \longrightarrow Mn$	-1.18	$Hg^{2+} + 2e^- \longrightarrow Hg$	0.80
$Zn^{2+} + 2e^- \longrightarrow Zn$	-0.76	$Ag^+ + e^- \longrightarrow Ag$	0.80
$Ga^{3+} + 3e^- \longrightarrow Ga$	-0.52	$Ir^{3+} + 3e^- \longrightarrow Ir$	1.00
$Fe^{2+} + 2e^- \longrightarrow Fe$	-0.44	$Br_2(l) + 2e^- \longrightarrow 2Br^-$	1.07
$Cd^{2+} + 2e^- \longrightarrow Cd$	-0.40	$O_2 + 4H^+ + 4e^- \longrightarrow 2H_2O$	1.23
$PbSO_4(s) + 2e^- \longrightarrow Pb(s) + SO_4^{2-}$	-0.35	$PbO_2(s) + SO_4^{2-} + 4H^+ + 2e^- \longrightarrow$ $PbSO_4(s) + 2H_2O$	1.69
$Tl^+ + e^- \longrightarrow Tl$	-0.34	$F_2(g) + 2e^- \longrightarrow 2F^-$	2.87
$Co^{2+} + 2e^- \longrightarrow Co$	-0.27		

IF TWO HALF-REACTIONS ARE COUPLED TO MAKE A WHOLE REACTION, THE HALF-REACTION HIGHER ON THE TABLE RUNS RIGHT TO LEFT, AS AN OXIDATION, AND THE LOWER HALF-REACTION IS THE REDUCTION. THE WHOLE REACTION'S VOLTAGE IS

$$\Delta E^0 = E^0 \textbf{(lower)} - E^0 \textbf{(higher)}$$

ΔE^0 IS ALWAYS A POSITIVE NUMBER!

Example: Lead-Acid Battery.

IN THE BATTERY UNDER
YOUR CAR'S HOOD, THE
ANODE IS METALLIC LEAD,
Pb(0), OXIDATION NUM-
BER 0. THE CATHODE IS
Pb(+IV), IN THE FORM OF
PbO_2. THE ELECTRODES
ARE IMMERSED IN STRONG
(6M) SULFURIC ACID, H_2SO_4.
THE OXIDATION AND
REDUCTION CHANGE BOTH
ANODE AND CATHODE
INTO Pb(+II).

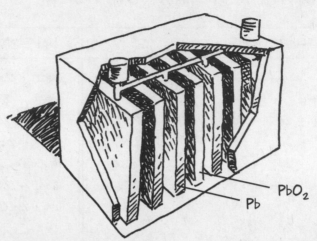

PbO_2

Pb

THE HALF REACTIONS ARE

OX: $Pb(s) + SO_4^{2-}(aq) \longrightarrow PbSO_4(s) + 2e^-$ $E^0_{RED} = -0.35\,V$

RED: $PbO_2(s) + SO_4^{2-}(aq) + 4H^+(aq) + 2e^- \longrightarrow PbSO_4(s) + 2H_2O$ $E^0_{RED} = 1.69\,V$

THE OVERALL REACTION ADDS UP TO

$$Pb(s) + PbO_2(s) + 2SO_4^{2-}(aq) + 4H^+(aq) \longrightarrow 2PbSO_4(s) + 2H_2O\,(l)$$

$$\Delta E = 1.69 - (-0.35) = \mathbf{2.04\,V}$$

CAR BATTERIES USUALLY PUT SIX OF THESE CELLS TOGETHER TO ACHIEVE A TOTAL
VOLTAGE OF 12V.

LEAD SULFATE IS INSOLUBLE AND BUILDS UP ON THE ELECTRODES WHILE SULFURIC
ACID AND THE ELECTRODES ARE CONSUMED. VOLTAGE DROPS...

BUT WHEN THE CAR IS
RUNNING, THE ENGINE'S
MOTION IS CONVER-
TED TO ELECTRICAL
ENERGY BY THE
ALTERNATOR. THIS
PUSHES ELECTRONS
BACK TOWARD THE
BATTERY'S ANODE, AND
THE REACTIONS ARE
REVERSED. THE BAT-
TERY **RECHARGES!**

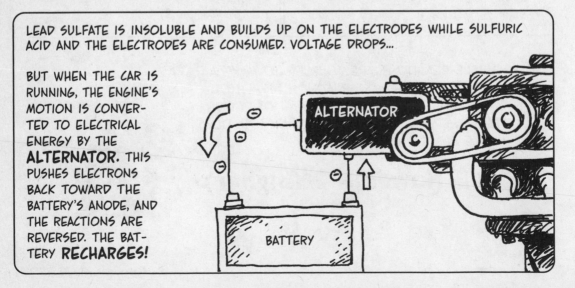

ALTERNATOR

BATTERY

Example: Fuel Cell

A FUEL CELL EXTRACTS ELECTRICAL ENERGY FROM A COMBUSTION REACTION SUCH AS

$$2H_2 + O_2 \longrightarrow 2H_2O$$

ONE KIND OF FUEL CELL INTRODUCES HYDROGEN AND OXYGEN ON OPPOSITE SIDES OF A POLYMER (PLASTIC) MEMBRANE. PROTONS CAN PASS THROUGH THE MEMBRANE, BUT IT BLOCKS ELECTRONS.

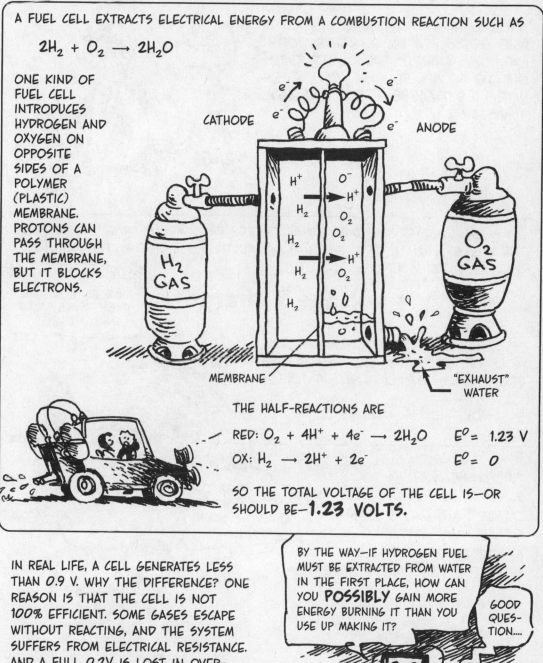

CATHODE

ANODE

H^+ H_2 H_2 H_2 H_2

O^- H^+ O_2 O_2 H^+ O_2

H_2 GAS

O_2 GAS

MEMBRANE

"EXHAUST" WATER

THE HALF-REACTIONS ARE

RED: $O_2 + 4H^+ + 4e^- \longrightarrow 2H_2O$ $E^0 = 1.23$ V

OX: $H_2 \longrightarrow 2H^+ + 2e^-$ $E^0 = 0$

SO THE TOTAL VOLTAGE OF THE CELL IS—OR SHOULD BE—**1.23 VOLTS.**

IN REAL LIFE, A CELL GENERATES LESS THAN 0.9 V. WHY THE DIFFERENCE? ONE REASON IS THAT THE CELL IS NOT 100% EFFICIENT. SOME GASES ESCAPE WITHOUT REACTING, AND THE SYSTEM SUFFERS FROM ELECTRICAL RESISTANCE. AND A FULL 0.2V IS LOST IN OVERCOMING THE REACTION'S **ACTIVATION ENERGY BARRIER.**

BY THE WAY—IF HYDROGEN FUEL MUST BE EXTRACTED FROM WATER IN THE FIRST PLACE, HOW CAN YOU **POSSIBLY** GAIN MORE ENERGY BURNING IT THAN YOU USE UP MAKING IT?

GOOD QUESTION....

Voltage and Free Energy

CAN WE PREDICT THE CHANGE IN VOLTAGE WHEN PRESSURES OR CONCENTRATIONS ARE NOT STANDARD? THE ANSWER TURNS OUT TO BE YES, BECAUSE VOLTAGE IS NOTHING BUT **GIBBS FREE ENERGY** IN DISGUISE.

EEK! MATH!

ON P. 213, VOLTAGE WAS DEFINED AS ENERGY DROP PER CHARGE, SO TO FIND THE ENERGY CHANGE OF A REACTION, WE MULTIPLY VOLTAGE BY THE AMOUNT OF CHARGE TRANSFERRED:

energy = voltage x charge

SPECIFICALLY, IF ONE VOLT MOVES ONE MOLE OF ELECTRONS, THE TOTAL ENERGY DROP TURNS OUT TO BE 96,485 JOULES.*

1 VOLT-MOL e^- = 96,485 J

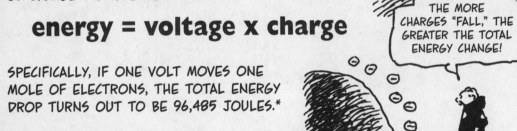

THE MORE CHARGES "FALL," THE GREATER THE TOTAL ENERGY CHANGE!

THIS CONVERSION FACTOR, 96.485 kJ/(VOLT-MOL e^-), IS CALLED **FARADAY'S CONSTANT,** AND WRITTEN $\mathscr{F}$. IF A VOLTAGE OF ΔE MOVES n MOLES OF ELECTRONS, THEN

ENERGY DROP = $n\mathscr{F}\Delta E$

THIS REPRESENTS THE MAXIMUM AMOUNT OF WORK THE CELL CAN POTENTIALLY DO.

HMM... "MAXIMUM WORK IT CAN DO..." SOUNDS FAMILIAR...

WHEEP!

*OBVIOUSLY, THE PERSON WHO DEFINED THE VOLT DIDN'T CONSULT WITH ANY CHEMISTS, WHO WOULD PROBABLY PREFER TO MEASURE ΔE IN UNITS OF 1/96,485 VOLT, OR "JOLTS" AND GET RID OF $\mathscr{F}$.

NOW THE MAXIMUM WORK A **REACTION** CAN DO IS −ΔG, WHERE ΔG IS ITS FREE ENERGY. AND A VOLTAIC CELL IS REALLY A REDOX REACTION! IN OTHER WORDS,

$$\Delta G = -n\mathscr{F}\Delta E \text{ JOULES, OR}$$

$$\Delta E = \frac{-\Delta G}{n\mathscr{F}} \text{ VOLTS}$$

LIKE, WHAT KIND OF WORK?

OH... LIKE TURNING OVER THE STARTER MOTOR, SAY...

THE MINUS SIGN IS AN ARTIFACT OF OUR DEFINITIONS. VOLTAGE IS THE SIZE OF THE ENERGY DROP, WHILE ΔG IS THE ENERGY CHANGE. SO ΔE > 0 WHEN ΔG < 0. THAT IS, **A REDOX REACTION IS SPONTANEOUS WHEN ΔE > 0.**

RUM RUM RUM RUM

RUM RUM RUM RUM RUM NER NER NER NER NER

IN THE LAST CHAPTER, WE SAW HOW ΔG CHANGES WITH CHANGING CONCENTRATIONS. IF WE HAVE A REACTION

$$aA + bB \rightleftharpoons cC + dD$$

THEN

$$\Delta G = \Delta G^0 + RT \ln Q$$

WHERE Q IS THE REACTION QUOTIENT

$$Q = \frac{[C]^c [D]^d}{[A]^a [B]^b}$$

SINCE $\Delta E = -\Delta G / n\mathcal{F}$ AT ANY CONCENTRATION, WE FIND

$$\Delta E = \Delta E^0 - (RT/n\mathcal{F}) \ln Q$$

THIS IS CALLED THE **NERNST EQUATION.** SINCE BALANCED HALF-REACTION POTENTIALS ARE REALLY WHOLE REACTION POTENTIALS MEASURED AGAINST A HYDROGEN ELECTRODE, THE EQUATION IS ALSO TRUE OF REDUCTION POTENTIALS E_{RED}.

$$E_{RED} = E^0_{RED} - (RT/n\mathcal{F}) \ln Q$$

AT EQUILIBRIUM, RECALL, $\Delta G = 0$, SO $\Delta E = 0$ AS WELL. THAT IS, WHEN $Q = K_{eq}$, THE BATTERY GOES DEAD.

CLICK
CLICK

YOU FORGOT TO PUT GAS IN THE CAR?

GRRRR...

THERE ARE MANY APPLICATIONS OF THE NERNST EQUATION. WE'LL LOOK AT ONLY ONE, WHEN pH = 7. (AT STANDARD CONDITIONS, REMEMBER, pH = 0!) pH 7 IS WHAT WE FIND IN LIVING ORGANISMS...

EXCEPT FOR CERTAIN SOUR INDIVIDUALS...

GAS

FOR SIMPLICITY'S SAKE, ASSUME H^+ APPEARS AS A **REACTANT** IN THE HALF REACTION (NOT A PRODUCT), AND ASSUME ALL OTHER SPECIES ARE AT STANDARD 1M CONCENTRATIONS OR CLOSE TO IT. IN THAT CASE WE WRITE THE ADJUSTED VOLTAGE AS $E^{0'}$.

$$E^{0'} = E^0 - (RT/n\mathcal{F})\ln Q$$

IF THE REACTION IS

$$hH^+ + aA + bB + \ldots \longrightarrow cC + dD + \ldots$$

AND $[A] = [B] = [C] = [D] = 1$. THEN **ALL FACTORS ARE EQUAL TO ONE** IN THE REACTION QUOTIENT, EXCEPT THE CONCENTRATION OF H^+!

$$Q = \frac{1}{10^{-7h}} = 10^{7h}$$

SO

$$E^{0'} = E^0 - (RT/n\mathcal{F})\ln(10^{7h})$$
$$= E^0 - (7hRT/n\mathcal{F})\ln(10)$$

BUT $\ln(10) = 2.3$, SO THIS

$$= E^0 - [(2.3)(7)hRT/n\mathcal{F}]$$

NOW ASSUME $h = n$, THAT IS, A MOLE OF HYDROGEN IS CONSUMED FOR EACH MOLE OF ELECTRONS, WHICH FREQUENTLY HAPPENS IN A NEUTRAL ENVIRONMENT. THEN PLUGGING IN ALL THE CONSTANTS GIVES THIS SIMPLE EQUATION:

$$E^{0'} = E^0 - 0.41 \text{ VOLTS!!!!!}$$

NOW WE CAN TALK ABOUT THE VOLTAGES WITHIN OUR OWN BODIES!

Glucose Oxidized

THE SUGAR GLUCOSE, $C_6H_{12}O_6$, IS THE BASIC FUEL OF LIFE AND A KEY INGREDIENT OF CELLS. IT OXIDIZES BY THIS EQUATION:

$$C_6H_{12}O_6 + 6O_2 \longrightarrow 6CO_2 + 6H_2O$$

THE HALF-REACTIONS ARE:

$$O_2 + 4H^+ + 4e^- \rightleftharpoons 2H_2O$$
$$6CO_2 + 24H^+ + 24e^- \rightleftharpoons C_6H_{12}O_6 + 6H_2O$$

(WRITTEN AS A REDUCTION AS ALWAYS!)

GIMME...

THE HALF-REACTIONS BOTH HAVE EQUAL AMOUNTS OF H^+ AND e^-, SO WE CAN USE THE FORMULA:

$$E^{0\prime} = E^0 - 0.41$$

OXYGEN'S REDUCTION REACTION IS IN THE TABLE ON P. 217, AND WE CAN WRITE

$$E^{0\prime} = 1.23 - .41 = \mathbf{0.82}\ V$$

WE CALCULATE E^0 OF THE OXIDATION REACTION FROM FREE ENERGY TABLES.

SPECIES	G^0_F (kJ/MOL)
$C_6H_{12}O_6$ (aq)	−917.22
CO_2	−394.4
H_2O	−237.18

"DO I DARE TO EAT A PEACH?"

$$\Delta G^0 = (-917.22) + (6)(-237.18) - (6)(-394.4)$$
$$= 26.1\ kJ/mol$$

$$E^0 = -\Delta G^0/n\mathcal{F} = -26.1/[(24)(96.485)]$$
$$= -0.011\ V$$

$$E^{0\prime} = -0.011 - 0.41 = \mathbf{-0.42}\ V$$

THEN THE VOLTAGE DROP FOR THE WHOLE REACTION IS

$$\Delta E^{0'} = E^{0'}(RED) - E^{0'}(OX)$$

$$= 0.82 - (-0.42)$$

$$= \mathbf{1.24} \textbf{ VOLTS} > 0$$

THE OXIDATION OF GLUCOSE IS SPONTANEOUS!!

WHICH RAISES THE QUESTION: **WHY DON'T WE ALL JUST BURST INTO FLAMES?** THE REASSURING ANSWER IS THAT SPONTANEOUS COMBUSTION IS STOPPED BY THE REACTION'S **ACTIVATION ENERGY.**

SO FAR THIS CHAPTER, WE'VE DESCRIBED HOW TO GET ELECTRICITY OUT OF A CHEMICAL REACTION... BUT WE HAVEN'T DISCUSSED HOW TO GET A CHEMICAL REACTION FROM ELECTRICITY.

ELECTROLYSIS IS WHAT HAPPENS WHEN A SUBSTANCE SPLITS AS THE RESULT OF AN APPLIED ELECTRIC CURRENT.

ALUMINUM, FOR EXAMPLE, IS EXTRACTED FROM ITS ORE ELECTROLYTICALLY...

UNFORTUNATELY, WE DON'T HAVE ROOM FOR THE DETAILS... AND SO ELECTROLYSIS WILL HAVE TO BE LEFT FOR ANOTHER DAY, ALONG WITH A FEW OTHER TOPICS TO BE DESCRIBED IN THE FOLLOWING CHAPTER.

Chapter 12
Organic Chemistry

IT'S ALIVE... OR IS IT?

OF THE NINETY-TWO NATURALLY OCCURRING ELEMENTS, SOME HAVE COMMANDED MORE OF OUR ATTENTION THAN OTHERS: HYDROGEN, FOR ITS ROLE IN ACIDS; OXYGEN, FOR ITS REACTIVITY AND LOVE OF HYDROGEN; BUT ONLY ONE ELEMENT DESERVES ITS VERY OWN BRANCH OF CHEMISTRY: **CARBON.**

THANKS TO ITS FOUR OUTER ELECTRONS, CARBON ATOMS CAN BOND WITH EACH OTHER TO FORM LONG CHAINS, WITH OTHER ATOMS ATTACHED TO THE LEFTOVER ELECTRONS. THE SIMPLEST OF THESE CHAINS ARE THE **HYDRO-CARBONS**, WHICH CONTAIN NOTHING BUT CARBON AND HYDROGEN.

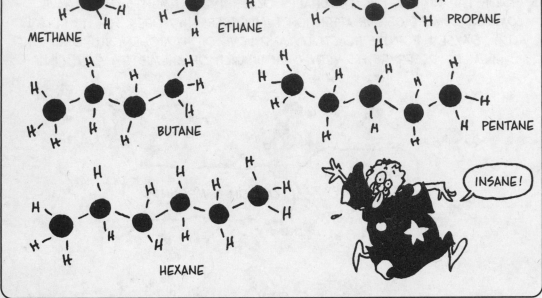

METHANE

ETHANE

PROPANE

BUTANE

PENTANE

HEXANE

INSANE!

CRUDE OIL IS MADE MAINLY OF HYDROCAR-BONS. SINCE LONG CHAINS HAVE HIGHER BOILING POINTS THAN SHORT ONES, OIL REFINERIES CAN SEPA-RATE ("FRACTIONATE") THEM BY LENGTH AND THEN CHEMICALLY "CRACK" THE LONG CHAINS INTO SHORTER ONES. GASOLINE IS A MIXTURE OF CHAINS WITH 5 - 10 CARBONS (OCTANE HAS 8).

HYDROCARBONS LIKE THOSE ON THE PREVIOUS PAGE, WITH SINGLE BONDS ONLY, ARE CALLED **ALKANES***. A DOUBLE BOND TURNS AN ALKANE INTO AN ALK**ENE,** AND A TRIPLE BOND MAKES IT AN ALK**YNE.** INDIVIDUAL MOLECULES ARE NAMED ACCORDINGLY.

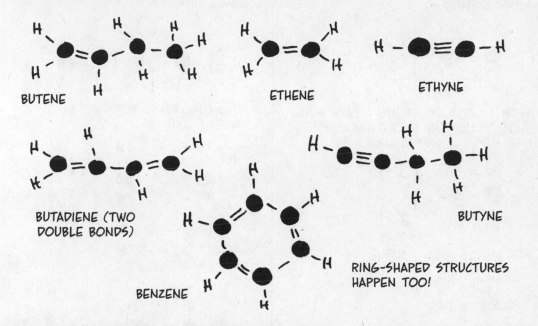

BUTENE

ETHENE

ETHYNE

BUTADIENE (TWO DOUBLE BONDS)

BENZENE

BUTYNE

RING-SHAPED STRUCTURES HAPPEN TOO!

TO COMPLICATE MATTERS FURTHER, TWO COMPOUNDS WITH THE SAME CHEMICAL FORMULA CAN HAVE DIFFERENT STRUCTURES. VARIANTS OF THE "SAME" MOLECULE ARE CALLED **ISOMERS.**

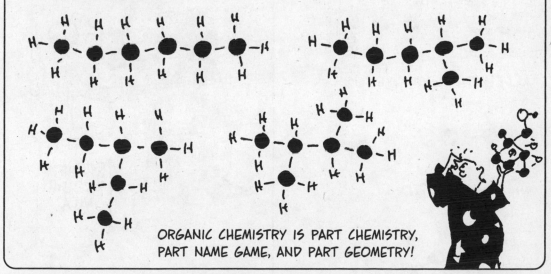

ORGANIC CHEMISTRY IS PART CHEMISTRY, PART NAME GAME, AND PART GEOMETRY!

*THEY ARE ALSO CALLED SATURATED HYDROCARBONS, SINCE THEY HAVE THE MAXIMUM POSSIBLE NUMBER OF HYDROGENS. ANYTHING WITH A DOUBLE OR TRIPLE BOND IS CALLED UNSATURATED.

IF A CHAIN HAS AN OH, IT'S CALLED AN **ALCOHOL.**

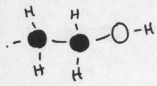

WITH A COOH GROUP, ITS A CARBOXYLIC **ACID.** (ONLY THE HYDROGEN COMES OFF, NOT THE WHOLE OH).

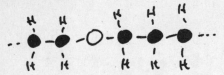

NH₂ MAKES IT AN **AMINE.**

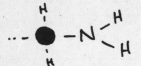

TWO CHAINS LINKED BY OXYGEN FORM AN **ETHER.**

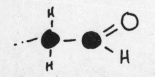

ALDEHYDES LOOK LIKE THIS:

AND THIS IS A **KETONE:**

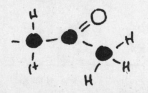

AND DON'T FORGET **ESTERS,** WHICH SMELL NICE.

THIS ONE, ETHYL FORMATE, SMELLS LIKE RUM...

AND PENTYL ACETATE IS "BANANA OIL."

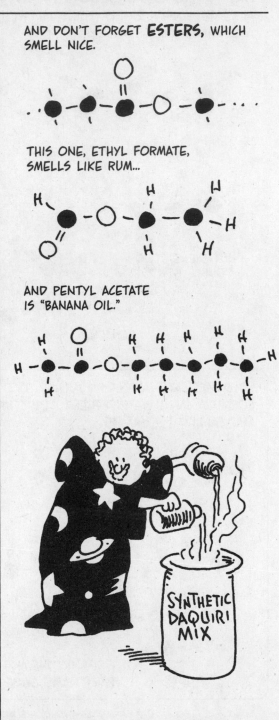

SYNTHETIC DAQUIRI MIX

CARBOHYDRATES ("HYDRATED CARBONS") HAVE EXACTLY TWICE AS MANY HYDROGENS AS OXYGENS.* THAT IS, THEIR GENERIC FORMULA IS $C_n(H_2O)_m$. THE SIMPLEST EXAMPLES ARE **SUGARS**, LIKE **GLUCOSE**, $C_6H_{12}O_6$.

ALPHA-GLUCOSE

BETA-GLUCOSE

HERE ARE THE TWO MAIN GLUCOSE ISOMERS. IN BETA, THE OH GROUP BESIDE O IS ON THE **SAME** SIDE OF THE RING AS THE SIDE CHAIN. IN ALPHA, OH IS ON THE **OPPOSITE** SIDE FROM THE CHAIN.

SINGLE-RING SUGARS ARE CALLED SIMPLE SUGARS OR **MONOSACCHA-RIDES.** SUCROSE, THE CANE SUGAR YOU BUY AT THE STORE, IS A **DISACCHARIDE** THAT LINKS ALPHA-GLUCOSE TO FRUC-TOSE, ANOTHER SIMPLE SUGAR.

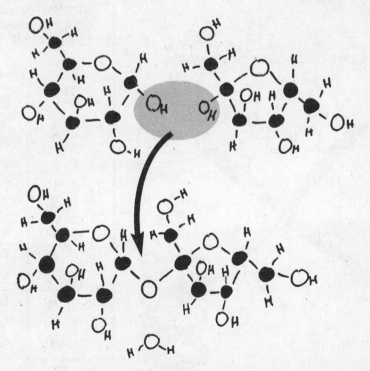

*THERE ARE EXCEPTIONS. DEOXYRIBOSE IS CONSIDERED A SUGAR, EVEN THOUGH IT IS ONE OXYGEN SHORT.

LET'S STOP A MOMENT AND ASK OURSELVES,

Why Carbon and Only Carbon?

WHY IS THIS THE ONE ELEMENT THAT FORMS LONG CHAINS?

6	
C	
12.01	
14	
Si	
28.09	
32	
Ge	
72.59	
50	
Sn	
118.7	
82	
Pb	
207.2	

SILICON, WHICH SITS BENEATH CARBON IN THE PERIODIC TABLE, ALSO HAS FOUR OUTER ELECTRONS, BUT WE DON'T SEE HYDROSILICON CHAINS.

ONE REASON IS THAT THE C–C BOND IS EXCEPTIONALLY STRONG. CARBON ATOMS ARE SMALL, SO THE SHARED ELECTRON CLOUD IS CLOSE TO THE NUCLEI, WHICH ATTRACT IT STRONGLY.

HERE ARE SOME BOND STRENGTHS OF INTEREST. (RECALL THAT THE NUMBERS MEAN THE AMOUNT OF ENERGY NEEDED TO BREAK THE BOND.)

NOR, FOR THAT MATTER, DO WE SEE CHAINS OF OXYGEN OR NITROGEN.

NEVER NEVER EVER!

BOND	STRENGTH(kJ/mol)
C–C	347-356*
C=C	611
C≡C	837
C–O	336
C–H	356-460*
Si–Si	230
Si–O	368
O–O	146
O=O	498
N–N	163
N=N	418
N≡N	946

*DEPENDING ON WHAT ELSE IS ATTACHED TO THE CARBON ATOM.

NOTE THAT THE C–C BOND IS EVEN STRONGER THAN THE C–O BOND. THIS MEANS THAT STABLE CARBON CHAINS CAN FORM IN THE PRESENCE OF OXYGEN.

YOU HAVE TO LIGHT A FIRE UNDER THEM TO GET THEIR ATTENTION!

BY CONTRAST, Si–Si BONDS ARE MUCH WEAKER THAN Si–O BONDS. OXYGEN DISRUPTS SILICON CHAINS. MOST SILICON ON EARTH EXISTS AS SiO_2 (SAND) OR SiO_3^{2-} IN SILICATE ROCKS. IN FACT, YOU OFTEN SEE OIL AND SAND SIDE BY SIDE.

SILICON'S EASY!

ALSO NOTE THAT TWO C–C BONDS ARE STRONGER THAN ONE C=C BOND. CARBON PREFERS THIS

TO THIS:

THREE SINGLE BONDS ARE ALSO STRONGER THAN ONE TRIPLE BOND. RESULT: LONG CHAINS ARE PREFERRED OVER SHORT ONES.

BY CONTRAST, OXYGEN PREFERS O=O TO O–O–O, AND NITROGEN PREFERS TO BOND WITH ITSELF AS N≡N. RESULT: NO OXYGEN OR NITROGEN CHAINS!

OTHERWISE, BREATHING WOULD BE TRULY **WEIRD!**

FINALLY, THE C–H BOND IS STRONG. HYDROCARBONS ARE STABLE AT ROOM TEMPERATURE. OTHER HYDRIDES TEND TO BE UNSTABLE AROUND OXYGEN.

YES, I'M SPECIAL!

 IN SUM, ONE OF CARBON'S PRE-FERRED STATES IS IN LONG SINGLY-BONDED CHAINS, POSSIBLY BRANCHED OR LOOPING BACK ON THEMSELVES AS RINGS, WITH A LOT OF HYDROGEN ATTACHED. **THIS IS TRUE OF NO OTHER ELEMENT.**

BIG, COMPLICATED CARBON MOLECULES FORM THE ESSENTIAL INGREDIENTS OF LIFE... IN FACT, CARBON COMPOUNDS ARE SO INTIMIATELY INVOLVED WITH LIVING SYSTEMS THAT CHEMISTS REFER TO ALL CARBON COMPOUNDS AS **ORGANIC.** CARBON MAKES LIFE POSSIBLE!

WE'D LIKE TO THANK THE ELEMENT CARBON...

LUCKILY FOR CHEMISTS, EVEN THE BIGGEST, MOST HORRIBLE ORGANIC COMPOUNDS ARE CHAINS OF SIMPLER SUBUNITS ATTACHED END TO END. THE SIMPLEST EXAMPLE IS POLYETHLENE PLASTIC, $(CH_2)_n$.

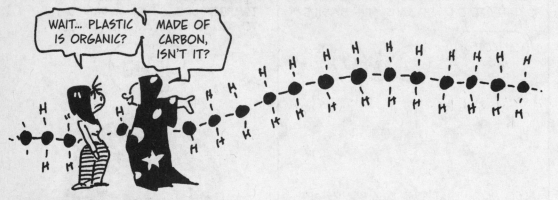

WAIT... PLASTIC IS ORGANIC?

MADE OF CARBON, ISN'T IT?

THE INDIVIDUAL UNITS OF THESE CHAINS ARE CALLED MONOMERS ("SINGLE TYPES"), AND THE WHOLE CHAIN IS A

polymer.

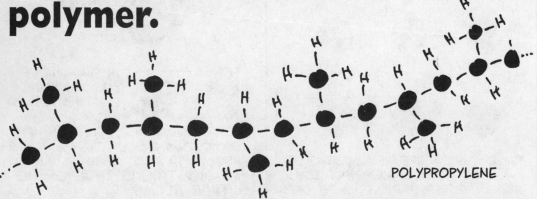

POLYPROPYLENE

NATURE'S POLYMERS ARE A BIT MORE WHIMSICAL THAN THESE SIMPLE PLASTICS. FOR INSTANCE, **POLYSACCHARIDES** COMBINE MANY SUGARS END TO END. **CELLULOSE** IS FORMED OF REPEATED UNITS OF BETA-GLUCOSE.

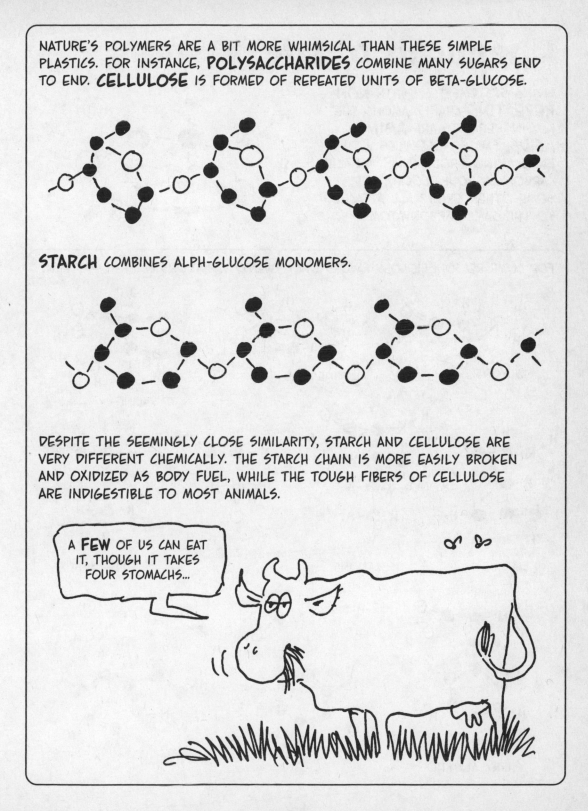

STARCH COMBINES ALPH-GLUCOSE MONOMERS.

DESPITE THE SEEMINGLY CLOSE SIMILARITY, STARCH AND CELLULOSE ARE VERY DIFFERENT CHEMICALLY. THE STARCH CHAIN IS MORE EASILY BROKEN AND OXIDIZED AS BODY FUEL, WHILE THE TOUGH FIBERS OF CELLULOSE ARE INDIGESTIBLE TO MOST ANIMALS.

A **FEW** OF US CAN EAT IT, THOUGH IT TAKES FOUR STOMACHS...

Chemicals of Life

LIVING SYSTEMS TEEM WITH **NON-REPEATING** CHAINS. AMONG THE KEY INGREDIENTS ARE **AMINO ACIDS,** SMALL MOLECULES WITH A BASIC AMINO GROUP (NH_2), AN ACID CARBOXYL GROUP (COOH), AND SOME OTHER GROUP ALL ATTACHED TO THE SAME CARBON ATOM.

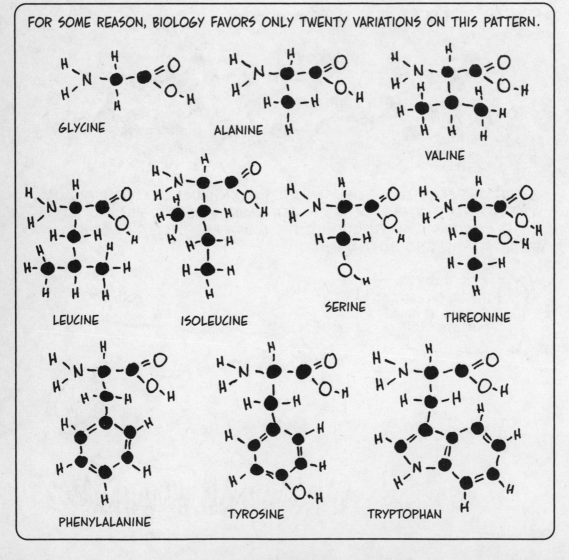

FOR SOME REASON, BIOLOGY FAVORS ONLY TWENTY VARIATIONS ON THIS PATTERN.

GLYCINE

ALANINE

VALINE

LEUCINE

ISOLEUCINE

SERINE

THREONINE

PHENYLALANINE

TYROSINE

TRYPTOPHAN

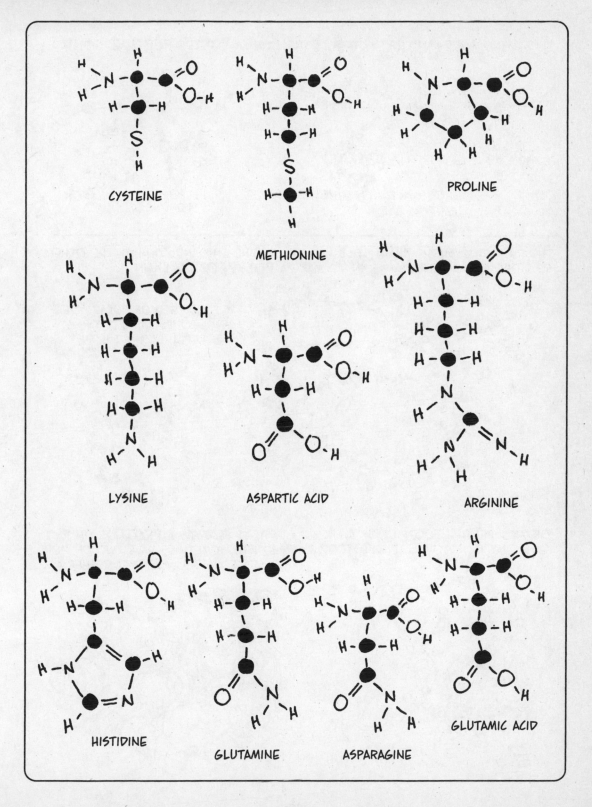

CYSTEINE

METHIONINE

PROLINE

LYSINE

ASPARTIC ACID

ARGININE

HISTIDINE

GLUTAMINE

ASPARAGINE

GLUTAMIC ACID

TWO AMINO ACIDS CAN LINK UP IN A CONNECTION CALLED THE **PEPTIDE BOND.**

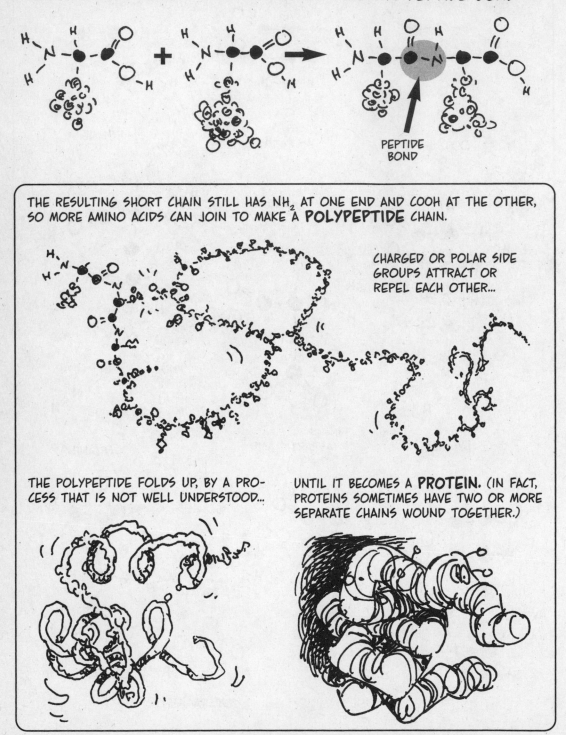

PEPTIDE
BOND

THE RESULTING SHORT CHAIN STILL HAS NH_2 AT ONE END AND COOH AT THE OTHER, SO MORE AMINO ACIDS CAN JOIN TO MAKE A **POLYPEPTIDE** CHAIN.

CHARGED OR POLAR SIDE GROUPS ATTRACT OR REPEL EACH OTHER...

THE POLYPEPTIDE FOLDS UP, BY A PROCESS THAT IS NOT WELL UNDERSTOOD...

UNTIL IT BECOMES A **PROTEIN.** (IN FACT, PROTEINS SOMETIMES HAVE TWO OR MORE SEPARATE CHAINS WOUND TOGETHER.)

SOME PROTEINS SERVE AS STRUCTURAL MATERIAL, BUT MOST ARE **CATALYSTS** FOR OTHER REACTIONS. CATALYTIC PROTEINS ARE CALLED **ENZYMES.** FOR EXAMPLE:

WHEN YOU EAT SUGAR, YOUR BODY MAKES ENZYMES THAT BREAK SUGAR DOWN...

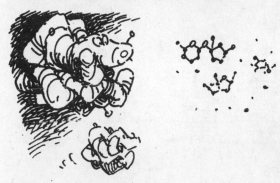

THE ENZYME RECOGNIZES THE PARTICULAR SUGAR MOLECULE...

AND CATALYZES THE REACTION THAT BREAKS IT DOWN INTO SMALLER PIECES.

THE ENZYME ITSELF IS UNCHANGED IN THE PROCESS.

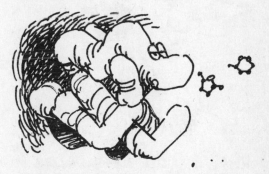

MEANWHILE, ANOTHER PROTEIN CALLED **HEMOGLOBIN** TRANSPORTS OXYGEN THROUGH THE BLOOD STREAM TO CELLS, WHERE IT CAN OXIDIZE GLUCOSE AND FREE THE ENERGY YOUR BODY NEEDS TO KEEP GOING.

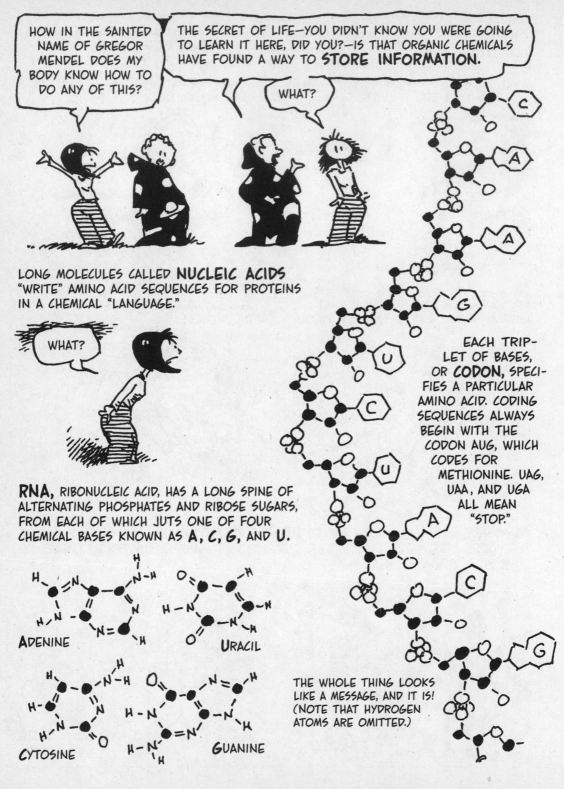

HOW IN THE SAINTED NAME OF GREGOR MENDEL DOES MY BODY KNOW HOW TO DO ANY OF THIS?

THE SECRET OF LIFE—YOU DIDN'T KNOW YOU WERE GOING TO LEARN IT HERE, DID YOU?—IS THAT ORGANIC CHEMICALS HAVE FOUND A WAY TO **STORE INFORMATION.**

WHAT?

LONG MOLECULES CALLED **NUCLEIC ACIDS** "WRITE" AMINO ACID SEQUENCES FOR PROTEINS IN A CHEMICAL "LANGUAGE."

WHAT?

RNA, RIBONUCLEIC ACID, HAS A LONG SPINE OF ALTERNATING PHOSPHATES AND RIBOSE SUGARS, FROM EACH OF WHICH JUTS ONE OF FOUR CHEMICAL BASES KNOWN AS **A, C, G,** AND **U.**

ADENINE

URACIL

CYTOSINE

GUANINE

EACH TRIPLET OF BASES, OR **CODON,** SPECIFIES A PARTICULAR AMINO ACID. CODING SEQUENCES ALWAYS BEGIN WITH THE CODON AUG, WHICH CODES FOR METHIONINE. UAG, UAA, AND UGA ALL MEAN "STOP."

THE WHOLE THING LOOKS LIKE A MESSAGE, AND IT IS! (NOTE THAT HYDROGEN ATOMS ARE OMITTED.)

THE OTHER NUCLEIC ACID, **DNA,** DEOXYRIBO-
NUCLEIC ACID, HAS TWO STRANDS SIMILAR TO
RNA'S WOUND AROUND EACH OTHER. LIKE
RNA, DNA USES THE BASES **A, C,** AND **G,**
BUT SUBSTITUTES **T** (THYMINE) FOR **U.**

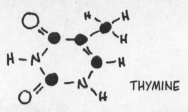

THYMINE

THE TWO STRANDS FIT TOGETHER WITH
MIRACULOUS PERFECTION: **A** ALWAYS PAIRS
WITH **T,** AND **C** ALWAYS PAIRS WITH **G,**
HELD TOGETHER BY HYDROGEN BONDS.

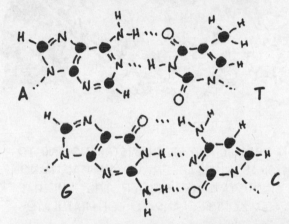

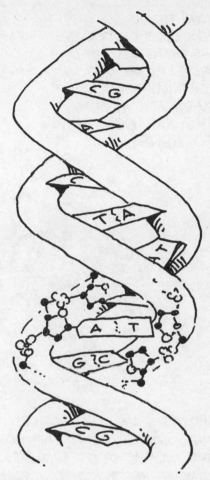

ONE STRAND OF DNA IS THE **COMPLEMENT** OF THE OTHER. IN OTHER WORDS,
DNA CARRIES THE INFORMATION NECESSARY TO **REPRODUCE ITSELF!!!**
(THE ACTUAL WORK IS DONE BY ENZYMES POWERED BY REDOX REACTIONS.)

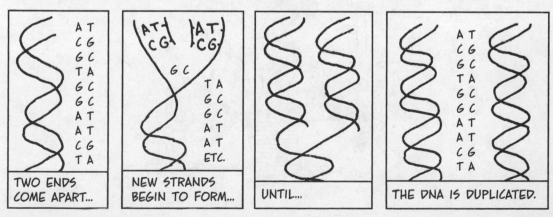

TWO ENDS COME APART...

NEW STRANDS BEGIN TO FORM...

UNTIL...

THE DNA IS DUPLICATED.

241

HOW IT DOES SO, AND HOW CODON SEQUENCES ARE TRANSLATED INTO PROTEINS, ARE DETAILS YOU WILL HAVE TO FIND ELSEWHERE. WE SUGGEST **THE CARTOON GUIDE TO GENETICS...**

AND THERE ARE A **LOT** OF DETAILS IN ORGANIC AND BIOCHEMISTRY, NO END TO THEM, IN FACT! NOT TO MENTION PHYSICAL, NUCLEAR, ENVIRONMENTAL, NANO-, AND ALL THE OTHER BRANCHES OF CHEMISTRY. YES, READER, THE TIME HAS COME TO REFER YOU TO MORE ADVANCED COURSES, AND TO CONGRATULATE YOU FOR GETTING THROUGH THE BASICS! 'BYE!

Appendix
Using Logarithms

IN SOME OF OUR CHAPTERS, WE USE A MATHEMATICAL SHORTHAND CALLED LOGARITHMS (OR LOGS, FOR SHORT). THE LOGARITHM IS A CONVENIENT, COMPACT WAY OF WRITING A NUMBER. FOR INSTANCE, INSTEAD OF $[H^+] = 10^{-7}$, WE WRITE pH = 7. pH IS A LOGARITHM.

I JUST HATE WASTING PENCILS...

A LOGARITHM IS AN EXPONENT. THE **COMMON LOGARITHM** OF A NUMBER N, log N, IS THE EXPONENT TO WHICH 10 MUST BE RAISED IN ORDER TO EQUAL N:

$$10^a = N \text{ IS THE SAME AS } a = \log N \quad \text{THAT IS, } 10^{\log N} = N$$

SO log 10 = 1 AND log 1 = 0 AND log 100 = 2 (SINCE $10^0 = 1$, $10^2 = 100$).

AND log 72.3 = 1.85914 BECAUSE $10^{1.85914} = 72.3$ (CHECK IT ON YOUR CALCULATOR.)

USED OFTEN IN CHAPTER 9!

KEY FACT: WHEN NUMBERS ARE **MULTIPLIED,** THEIR **LOGARITHMS ARE ADDED.**

$$\log MN = \log M + \log N$$

THIS IS BECAUSE $10^a 10^b = 10^{(a+b)}$. IF $M = 10^a$ AND $N = 10^b$, THEN $MN = 10^a 10^b = 10^{(a+b)}$, SO $a+b = \log MN$. BUT $a = \log M$ AND $b = \log N$.

SIMILARLY

$$\log(M^p) = p(\log M)$$
$$\log\left(\frac{1}{N}\right) = -\log N$$

BECAUSE THIS IS HOW EXPONENTS BEHAVE:

$$10^{-a} = \frac{1}{10^a} \qquad 10^{ab} = (10^a)^b$$

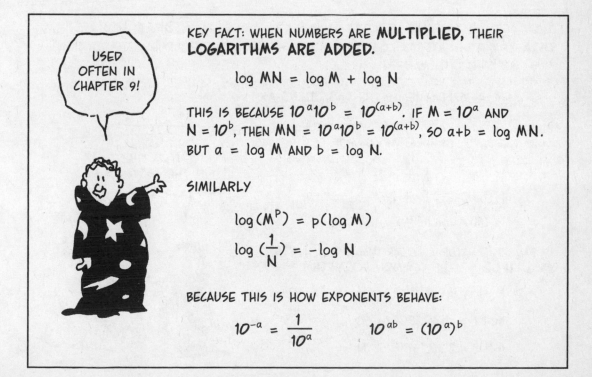

log N GIVES US A ROUGH IDEA HOW BIG N IS. THE WHOLE-NUMBER PART OF THE LOGARITHM GIVES N'S ORDER OF MAGNITUDE.

$$\log 1{,}234 = 3.0913$$
$$\log 1.234 = 0.0913$$
$$\log 1{,}234{,}000 = 6.0913$$
$$\log (a \times 10^n) = n + \log a$$

THERE'S A GOOD ONLINE CALCULATOR AT http://www.squarebox.co.uk/desktop/scalc.html

WHAT A HUGE NUMBER...

Natural Logarithms

COMMON LOGS HAVE BASE TEN. THEY ARE EXPONENTS OF 10. SOMETIMES, THOUGH, THEY ARE LESS CONVENIENT THAN "NATURAL LOGS," FOR INSTANCE, WHEN A QUANTITY CHANGES AT A RATE PROPORTIONAL TO ITSELF. THAT IS, AT TIME t,

$$r_A(t) = kA_t$$

THEN IT'S POSSIBLE TO SHOW THAT THE QUANTITY A_t IS AT ANY TIME t IS

$$A_t = A_0 e^{kt}$$ WHERE A_0 IS THE INITIAL AMOUNT OF A, AND $e = 2.71828$

THEN $e^{kt} = A_t/A_0$ AND WE WRITE $kt = \ln(A_t/A_0)$, THE **NATURAL LOGARITHM** OF A_t/A_0. THE NATURAL LOG OF ANY NUMBER N IS THE EXPONENT TO WHICH e MUST BE RAISED TO MAKE N.

$$M = \ln N$$ MEANS THE SAME THING AS $e^M = N$

BECAUSE $e^a e^b = e^{(a+b)}$, ETC., THE NATURAL LOGS OBEY THE SAME FORMULAS AS COMMON LOGARITHMS.

$$\ln MN = \ln M + \ln N$$
$$\ln(1/M) = -\ln M$$
$$\ln(M^n) = n \ln M$$

JOHN NAPIER INVENTED THESE IN THE 1600s.

IN FACT, THE NATURAL LOGARITHM IS A **CONSTANT MULTIPLE** OF THE COMMON LOGARITHM.

$$\ln N = \ln(10^{\log N}) = (\log N)(\ln 10)$$
$$\ln 10 = 2.302585..., \text{ SO}$$
$$\ln N = 2.302585 \log N$$

Index

absolute entropy, 196–97
acids and bases, 165–90
  buffers, 185–89, 190
  conjugate pairs, 166, 167, 186
  equivalent weight of, 178
  neutralization, 177–80
activation energy, 151–54, 219, 225
air, 4, 10, 98
alchemy, 5–6
alternator, 218
amino acids, 236–38, 240
ammonia, 59, 163, 167, 176, 179
amu (atomic mass unit), 25, 72
anions, 20, 41, 43, 50, 212
  single-atom, 48
anode, 19, 212, 213, 218
Aristotle, 4–5, 11
atmospheric pressure, 7–8, 111, 142
atomic mass, 24–26, 28
atomic number, 25–27, 40
atomic size, 39
atomic weight, 11, 12, 15, 26, 112
atoms, 4, 13
  atomic theory, 19–44
  atomists, 4, 13
  atom building, 34–39
  bonds between, 45–66
  electron affinity, 41–44
  electronegativity, 47, 48, 54, 56, 62, 63
  ionization energy, 40
  net charge, 78
  oxidation number, 79, 210
  See also electrons
attractions, 106–28

Avogadro's law, 112
Avogadro's number, 72

balanced equations, 70–73, 81
bases. See acids and bases
battery, 19, 213, 218, 222
boiling point, 109, 119–21
  carbon chains, 228
  dissolved material, 139
  heating curve, 126–27
  helium, 125
bomb calorimeter, 96
bonds, 45–66
  carbon atoms, 228, 232
  potential energy in, 87
  solvation, 131–32
  strength of, 108, 232–33
  See also intermolecular forces
Boyle's law, 112
Brand, Hennig, 5
buffers, 185–89, 190
bystander ion, 180

calorimetry, 96–100
carbohydrates, 231
carbon, 14, 34, 47, 82, 227, 232–233
  atom, 21, 24, 25, 228
  hybrid orbital, 60
  oxidants/reductants, 80–81
  phase diagram, 125
  valence electrons bonds, 58
carbon chains, 228–41
catalysts, 153–54, 239

catalytic converter, 154
cathodes, 19, 20, 212, 213
cations, 20, 182, 212
Celsius scale, 88
Charles's law, 112
chemical bonds. See bonds
chemical reactions, 8–12, 67–83
  activation energy, 151–54
  alchemy as, 5–6
  catalysts, 153–54, 239
  defined, 2
  electricity from, 209–26
  as energy transfer, 89–104
  entropy and, 198–206
  fire as first, 1–3
  free energy, 205
  higher-order, 155–57
  hydrolysis, 175
  rate of, 141–64
  redox, 76–77
  reversible, 158–59, 195, 207
  solutions and, 129–40
  spontaneous, 201
collision theory, 146–52
combination reaction, 69, 146–52
combustion, 11, 68, 69, 77, 219
  heat of, 103
  spontaneous, 225
compounds, 11–13, 79, 229
concentration, 133–34, 142–43, 164, 168–69, 182
condensation, 118–21
coolants, 94, 95, 117
copper, 3, 93–94
  zinc reaction, 14, 212–13
corrosion, 6, 77

covalent bond, 54–58, 62–63, 65
  strength of attraction, 108
crystalline structures, 48–51
  of carbon, 125
  covalent bonds, 57
  of ice, 123
  ionic bonds, 48–51, 64
  metallic bonds, 51, 52–53
current, electric, 19, 53, 226

Dalton, John, 13
decomposition reaction, 69
Democritus, 4
dipoles, 106–7
dissolving process, 129–40
  acids and bases, 168–69, 184
  freezing/boiling points, 138–39
  salts in water, 129, 130, 182
DNA, 241
double bond, 56, 58, 61
double-displacement reaction, 76
dynamic balance, 158–59

elasticity, 110
electric cells, 211, 212
electric potential, 213
electricity, 17–44, 209–26
  attractions/repulsions, 90, 106–28
  metal conductors, 53
  See also negative charge; positive charge
electrochemistry, 209–26
electrodes, 20, 212, 218
electrolysis, 19, 20, 226
electromagnetic radiation, 87
electronegativity, 47, 48, 54, 56, 62, 63

electrons, 20, 21, 24, 26, 28–44
  affinity, 41–44
  bonds, 47, 52, 54–58, 63, 232
  dipole attraction, 107
  ionization energy, 40
  metal, 52, 53
  orbit, 29–33, 36, 60
  outer, 39, 40, 56
  paired, 58–59, 61
  particle/wave, 28, 30
  redox reactions, 77–81, 103, 209–19
  rule of eight, 43–44, 61
  sharing, 57, 58–59
  shells, 31–39
electropositivity, 47, 48, 54, 62
electrostatic attraction, 48
elementary reactions, 156, 157
elements, 12–16
  ancient four, 4, 10, 11
  atomic number, 25
  carbon's uniqueness, 232–33
  charge extremes, 62
  grouping of, 36–37
  isotopes of, 25
  list of, 27
  oxidation number, 78, 79
  periodic table, 15–16, 38–44
empirical formula, 49, 68
emulsion, 132
endothermic reactions, 99, 102, 116, 122, 151
energy, 26, 30, 31, 39, 85–103
  activation, 151–54, 225
  collision, 150–51
  conservation law, 86
  electrical, 209–26
  quanta of, 30, 194
  spreading out of, 194, 195–202
  transfer of, 89–104
enthalpy, 98–99
  change, 131, 200, 201

of formation, 100–104, 116, 122, 205
entropy, 195–206
enzymes, 239
equilibrium, 118, 124, 158–64, 201, 222
  acids and bases, 165–90
equilibrium constant, 160–61, 175, 182
  pH, 170
  second derivation of, 207–8
  solubility product, 182–84
  weak ionization, 172–73
equivalent weight, 178
evaporation, 116–19, 122, 126–28, 139
exothermic reactions, 99, 104, 151
explosions, 98, 99, 102–3, 114
explosives, 6, 76–77, 80–83

Faraday's constant, 220
fire, 1–3, 4, 9, 11, 67, 68
first-order reaction, 145
forward reaction, 159, 182, 199, 207
four basic elements, 4, 10, 11
Franklin, Benjamin, 18
free energy change, 201–6, 220–23
free radical, 142
freezing point, 95, 123, 138
fuel cell, 219

gases, 6–13, 98, 110–14
  characteristics of, 105
  noble, 43–44, 107, 125
  solubility, 137
  state changes, 116, 121, 124–25
  temperature and, 91, 109

gas laws, 112–14, 128
Gibbs function, 201–5, 220
Gilbert, William, 17
glucose, 213, 224–25, 239
Guericke, Otto von, 7, 111
gunpowder recipe, 82

Haber process, 163, 200, 204
half-life, 143–44
half-reactions, 214–19, 222, 224
halogens, 41
heat, 86–104
    reaction activation, 151–54
    See also temperature
heat capacity, 92–97, 197
heat change, 93, 96–104, 200
heating curves, 126–28
heat of combustion, 103
heat of fusion, 122
heats of formation, 100–104
helium, 125
hemoglobin, 239
Henderson-Hasselbalch equation, 187–89
Heraclitus, 4
Hess's Law, 101
Higher-order reactions, 155–57
hybrid orbitals, 60
hydrocarbons, 228–30, 233
hydrogen, 9, 12, 13, 214, 227
    atomic number, 26
    carbon chains, 228–31, 233
    electron shell, 31, 34, 56
    heat of combustion, 103
    pH, 171
    positive charge, 19, 62
    redox reaction, 214
hydrogen bond, 55, 64, 94, 106
    attraction strength, 108, 109
    DNA, 241

hydrolysis, 175
hydronium, 168

ice, 123, 126–27
ideal gas, 110, 113
in solution, 130, 134, 161
indicator chemicals, 171
intermolecular forces, 106–9
internal energy, 90–91
ion, 20, 31, 48, 49, 51, 109
ionic bonds, 48–51, 54, 65
    dipole, 106–8
    polarity, 63
    strength of attraction, 108
ionic crystals, 48–51
ionic repulsion, 51, 53
ionization, 31, 40
    base constant, 175–76
    equilibrium, 160–64
    high, 43
    ionization energy, 40
    of water, 161, 168, 170, 172, 185–89, 208
    weak, 172–76
isomers, 229
isotopes, 25

Jabir, 5
Joule, James Prescott, 92
Joules, 86, 92, 93, 127

Kelvin scale, 88, 110
kinetic energy, 87, 90–91, 150

lanthanide series, 37
Lavoisier, Antoine, 10–11
Lead-acid battery, 218, 222

Le Chatelier's principle, 162–63, 184, 204
Lewis diagram, 56, 59, 61
life
    chemicals of, 236–41
    glucose oxidation, 224–25
    hydrogen bonding, 64
    origin of, 154
liquids, 105, 106, 109, 115–21
    boiling point, 119–20
    evaporation/condensation, 116–21, 122
    melting point, 123
    phase diagrams, 125–26
    solubility, 135–37
    solutions, 129–40
    standard molar energy, 197
    surface tension, 115
    suspensions, 132
    See also water
logarithms, 171, 243–44
London dispersion force, 107

main-group elements, 37
mass, 24, 28, 72
mass action, law of, 160
mass-balance table, 73, 82
matter, 2–44, 105–28
    ancient theories of, 4–5, 13
    three types of, 105
mechanical energy, 87
melting point, 109, 122–23
    heating curve, 126–27
Mendeleev, Dmitri, 15
metal ions as acids, 173
metallic bonds, 52–53, 108
metals, 42, 211
miscibility, 135
molar heat capacity, 92
molarity, 134
mole, 72–73, 81, 110, 112
    Avogadro's number, 72

molecules, 13, 49, 55–61
  attractions between, 106–9
  charged, 61, 63
  collision theory, 146–52
  composition, 57
  ionization fraction, 174
  kinetic energy storage, 194
  shapes, 58–59
  solubility, 136, 139
  standard entropy, 197
  weight, 72
mullite, 69, 70

negative charge, 18–22, 28, 212
  electron, 20, 24
negative reduction potential, 217
neon, 34, 43
Nernst equation, 222, 223
neutralization, 177–81, 190
neutrons, 24, 25, 26
noble gases, 43–44, 107, 125
non-metals, 42, 47, 56
nonrepeating chains, 236–38
nucleic acids, 240–41
nucleus, 22, 25–28, 41

orbitals, 29–36, 43, 60
organic chemistry, 227–42
oxidants, 80, 103
oxidation, 77, 224–25
oxidation numbers, 78–83, 210
oxidation-reduction. See redox reactions
oxygen, 9–14, 47, 227, 239
  atomic number, 26
  carbon chains, 230, 231, 233
  covalent bond, 56, 58

electron shells, 34
  negative charge, 19, 62
ozone, 142

partial pressure, 118, 119, 122, 137, 146–48
particles, 20, 24, 28, 48
  collision of, 146–52
  entropy, 198
  number in mole, 72
peptide bond, 238
periodic table, 15–16, 38–44
pH, 170–71, 173, 176, 178–80
  buffers, 185–89
  endpoint, 181
  Nernst equation, 223
  solubility effects, 184
phase change, 109, 119–27, 195
phase diagrams, 124–25
photons, 87
picometer, 22
plasma, 128
polarity, 62–65, 136
polyatomic atoms, 50, 61, 78
polymers, 234–35
polypeptide chain, 238
positive charge, 18–22, 28, 212
  proton, 24
potential energy, 87, 90, 213
pottery, 69, 70, 73, 117
precipitating, 68
pressure, 110–12, 124
  constant, 98, 99
  entropy change, 206
  external, 119–20, 123
  gas law equation, 113, 122
  gas solubility, 137
  ice melting, 123
  Le Chatelier's principle, 163, 204
  vapor, 118–22, 139

Priestley, Joseph, 8–9, 11
properties, 1–16, 54
  metals vs. nonmetals, 42
proteins, 238–39, 240
protons, 24–27

quantized energy, 30, 194
quantum mechanics, 28, 29, 61, 198

radiant energy, 86, 87
rate constant, 144
Razi, al-, 5
reactants, 68–69, 141–64, 202, 223
  enthalpy of formation, 101, 116
  mass-balance table, 73
  See also chemical reactions
reaction constant, 153–54
reaction equations, 68, 73, 143–45, 207
reaction products, 68
reaction quotient, 207
reaction rate, 141–64
reaction stoichiometry, 71
redox reactions, 76–83, 103, 209–21
reductants, 80
resonance, 61
reverse reaction, 158–59, 195, 207
RNA, 240
rule of eight, 44, 61

salt, 20, 41, 48, 51
  acid-base neutralization, 177–80, 190
  boiling point, 139

dissociation in liquid, 64, 129, 130, 182
solubility products, 182–83
saturation, 135, 182–84
second-order reactions, 146–47, 153–55
soap, 75
solids, 105, 106, 109, 122–26
dissolved, 130–32
standard molar entropy, 197
solubility, 135–37, 184
products, 182–83
solutions, 129–40
acidity measure, 168–76
buffers, 185–89
neutralization, 178–80
pH, 171, 178–80
reaction rate, 142–48
saturation, 182–84
titration, 181
weak acid, 174–76
solvation, 131–32, 138–39
specific heat, 92, 93–95, 127
spontaneous processes, 192–93, 201, 204, 221, 225
starch, 235
stoichiometric coefficients, 160
sublimation, 122, 124
sugars/sucrose, 130, 231, 239

superfluid, 125
surface tension, 115
suspensions, 132

temperature, 88–89, 91, 104
boiling point, 120
calorimetery, 96–97
critical, 121
entropy change, 195
gas law equation, 113
heat capacity, 92–95
melting point, 122–25
reaction rate, 152, 164, 204
solubility, 135, 137
state effects of, 109
thermodynamics, 191–208
second law of, 199
thermometers, 88, 115
titration, 181
transition metals, 37, 39
transition state, 149

valence electrons, 39, 40, 56, 58, 79
vapor pressure, 118–22, 139
vinegar, 130, 174

voltaic cell, 213
volts/voltage, 31, 213, 215–18, 225
free energy and, 220–23
volume, 110, 112, 113

water, 12, 13, 14, 19, 196
acids/bases, 168–69, 172, 185–89
boiling point, 119–20
dipole molecule, 106
evaporation, 116–17, 127, 177
freezing expansion, 123
ionization, 161, 168, 170, 172
ionization constant, 161, 170, 208
melting point, 123
molecular shape, 59
polarity, 62–63, 64
specific heat, 93, 94, 95, 127
splitting, 175
water constant, 170
wavelength, 28, 29, 30
weights, 11, 12, 15, 72, 178
work energy, 86, 98, 202, 221

About the Authors

LARRY GONICK IS THE SON AND SON-IN-LAW OF CHEMISTS. HE ONCE CONSIDERED A SCIENTIFIC CAREER, BUT WISELY ABANDONED THE IDEA AFTER BREAKING TWELVE PIECES OF GLASSWARE IN A SINGLE, DISTRESSING THREE-HOUR CHEMISTRY LAB. HE WRITES AND DRAWS NONFICTION COMIC BOOKS AND IS THE STAFF CARTOONIST FOR **MUSE** MAGAZINE. HE LIVES PHYSICALLY IN CALIFORNIA WITH HIS FAMILY AND VIRTUALLY ON THE WEB AT www.larrygonick.com.

CRAIG CRIDDLE IS PROFESSOR OF ENVIRONMENTAL ENGINEERING AND SCIENCE AT STANFORD UNIVERSITY, WHERE HE TEACHES AQUATIC CHEMISTRY AND ENVIRONMENTAL BIOTECHNOLOGY. HE HAS PUBLISHED MANY ARTICLES ON CHEMICALS IN WATER AND WATER CLEANUP, AND HIS TEAM OF GRAD STUDENTS AND RESEARCH ASSOCIATES LIKE TO THINK THEY CAN SOLVE THE WORLD'S WATER CRISIS. PROF. CRIDDLE AND HIS WIFE LIVE IN CUPERTINO, CALIFORNIA, ALONG WITH THEIR DOG AND WHICHEVER OF THEIR FOUR KIDS (MOSTLY GROWN) HAPPENS TO BE HOME. HIS WEB SITE IS www.stanford.edu/group/evpilot/. HE BELIEVES THAT BROKEN EQUIPMENT IS A NATURAL PART OF SCIENCE.